SCIENCE ACTIVITY LAB

DK | Penguin Random House

SECOND EDITION

Editor Jolyon Goddard

Managing editor Rachel Fox
Managing art editors Govind Mittal, Owen Peyton Jones
DTP coordinator Vishal Bhatia
Production Editor Anita Yadav
Production controller Jack Matts
Jacket designer Vidushi Chaudhry
Jacket design development manager Sophia MTT
Jackets editorial coordinator Priyanka Sharma Saddi
Jacket DTP designer Deepak Mittal

Publisher Andrew Macintyre
Associate publishing director Liz Wheeler
Art director Karen Self
Publishing director Jonathan Metcalf

Writer Jack Challoner

FIRST EDITION

Senior designers Michelle Staples, Jacqui Swan
Lead editor Amanda Wyatt
Editors Steven Carton, Ben Morgan
Designers Sean T. Ross, Chrissy Barnard, Alex Lloyd, Gregory McCarthy, Mary Sandberg
Illustrators Alex Lloyd, Sean T. Ross, Gus Scott

Managing editor Lisa Gillespie
Managing art editor Owen Peyton Jones
Producer, pre-production Gill Reid
Senior producer Meskerem Berhane
Jacket designers Tanya Mehrotra, Surabhi Wadhwa-Gandhi
Jacket design development manager Sophia MTT
Jackets editor Emma Dawson
Managing jackets editor Saloni Singh
Jackets editorial coordinator Priyanka Sharma
Jacket DTP designer Rakesh Kumar
Picture researcher Rituraj Singh

Publisher Andrew Macintyre
Associate publishing director Liz Wheeler
Art director Karen Self
Publishing director Jonathan Metcalf

Writer and consultant Jack Challoner
Photographer Dave King

This edition published in 2024
First published in Great Britain in 2019 by
Dorling Kindersley Limited
DK, One Embassy Gardens, 8 Viaduct Gardens,
London, SW11 7BW

Content previously published as *Science Lab*

The authorised representative in the EEA is
Dorling Kindersley Verlag GmbH. Arnulfstr. 124,
80636 Munich, Germany

Copyright © 2024, 2019 Dorling Kindersley Limited
A Penguin Random House Company
2 4 6 8 10 9 7 5 3 1
001–339264–October/2024

A CIP catalogue record for this book
is available from the British Library.
ISBN: 978-0-2416-5702-7

Printed and bound in China

www.dk.com

This book was made with Forest
Stewardship Council™ certified
paper – one small step in DK's
commitment to a sustainable future.
Learn more at **www.dk.com/uk/
information/sustainability**

ROBERT WINSTON

SCIENCE ACTIVITY  LAB

FANTASTIC PROJECTS FOR YOUNG SCIENTISTS

DK

CONTENTS

SCIENCE FACTS
This symbol points out facts about biology, chemistry, or physics.

TECHNOLOGY FACTS
This symbol highlights more information about tools or materials.

ENGINEERING FACTS
This symbol directs you to more facts about structures or machines.

MATHEMATICS FACTS
This symbol identifies extra information on formulas, shapes, or measurements.

FOREWORD

From the age of eight, I had huge fun building models. The sense of achievement I had when they occasionally worked is amongst my most treasured memories. They ranged from cranes to aeroplanes, models of famous buildings, or, on one occasion, a mechanical contraption for passing written messages across the ceiling of my classroom. Amazingly, it was a long time before my teacher tumbled to what we were doing in class. I once made a radio amplifier from cheap components. I turned on the power, there was a humming noise, and blue smoke filled the dining room!

Nearly all my models initially ended in failure. Failure is very important because if we use it properly, we learn from our mistakes and rebuild with great improvement. All scientists encounter failure regularly; only by embracing it do we eventually get our experiments to work.

Modelling is really important, especially for scientists, engineers, and architects. Nobody would dream of designing a structure like a bridge or a stadium without making models first. This helps them understand the science behind the structures, the mathematical principles involved, and how recalculation may be needed to achieve their final ambition. Even the simplest experiments in this book show how modelling can help resolve practical problems. How does the Air freshener (page 74) behave at lower temperatures, or high humidity, for example? Test yours and find out.

The models and projects in this book allow you to consider improvements yourself. For example, the sturdy Bottle raft (page 18) offers possibilities. It might even be powered by an elastic band and wooden paddles, by a sail and wind power, or by a tiny electric fan. When my daughter was eight years old, she and I built a fan-powered boat as a school project. We even raced it across the swimming pool! It was a huge success, and several versions were then made by her classmates.

You'll also want to think about materials. Materials scientists calculate what are the most practical substances to increase strength, for example, or reduce weight. What are the best materials for the blades of the Wind turbine (page 30) or the mainspring of the Wind-up car (page 10)? As you build the projects in this book, you may find alternative materials that help you to construct more effective models. And the more care you take in measuring the components precisely and drawing straight lines and accurate angles, the more pleasure you will get from seeing an effective working machine that you built yourself.

I hope you enjoy exploring science, technology, engineering, and maths in action as much as I have enjoyed doing so with my own models.

ROBERT WINSTON

FORCES AND MOTION

A force is a push or a pull, and there are forces at work everywhere! Forces can make things move or stop moving, make things speed up or slow down, or just keep things still. One of the most familiar forces is gravity, which pulls everything down towards the ground. In this chapter, you'll be fighting against gravity by constructing a crane and by making a ping-pong ball hover in the air. You'll also explore the forces that make a raft stay afloat.

STEM YOU WILL USE

• SCIENCE: Friction between a turning wheel and a surface is what drives a car along.

• ENGINEERING: Flexible materials, such as paper and card, can store useful energy as they bend.

The coiled paper mainspring stores the energy to power the car.

The axle is the rod that connects the wheels together.

The wheels are made of plastic bottle lids.

WIND-UP CAR

Used for centuries to make clocks and moving toys, wind-up mechanisms have long, coiled strips of springy material called mainsprings that store energy as they're tightened. Energy can't be created or destroyed, it can only be transferred. So as you wind up the car, its mainspring stores the energy you put into turning it. Let it go and VROOM! The energy is released and your car is off!

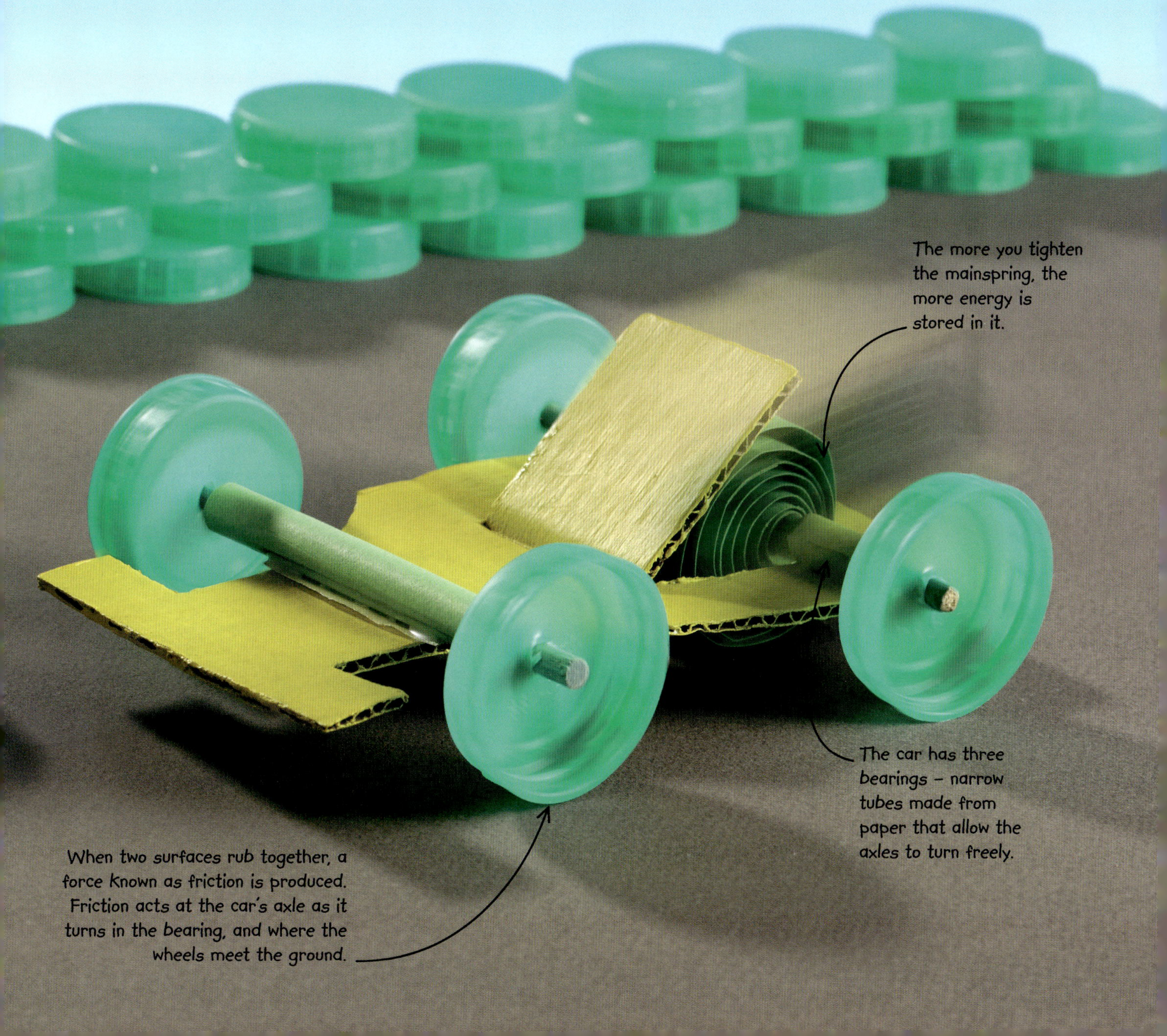

The more you tighten the mainspring, the more energy is stored in it.

The car has three bearings – narrow tubes made from paper that allow the axles to turn freely.

When two surfaces rub together, a force known as friction is produced. Friction acts at the car's axle as it turns in the bearing, and where the wheels meet the ground.

HOW TO MAKE A
WIND-UP CAR

This wind-up car is powered by energy stored in a coiled mainspring made of paper. Its axles (the rods connecting the wheels) are made from a garden stick, while its bearings (the tubes that allow the axles to turn freely) are made with paper. The axles and bearings are attached to the car's frame, or chassis.

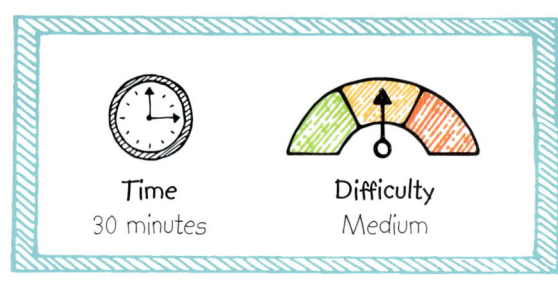

Time
30 minutes

Difficulty
Medium

WHAT YOU NEED

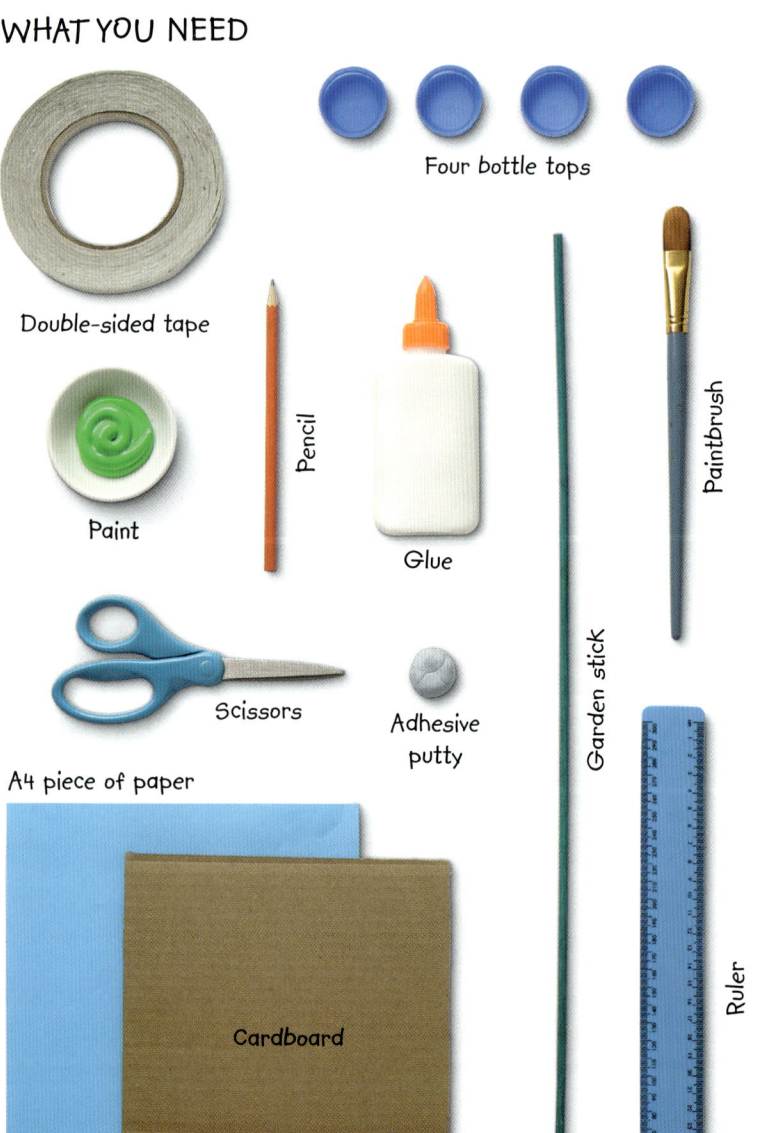

Double-sided tape

Four bottle tops

Paint

Pencil

Glue

Paintbrush

Scissors

Adhesive putty

Garden stick

A4 piece of paper

Cardboard

Ruler

1 Draw a rectangle 15 cm (6 in) long and 8 cm (3 in) wide on the cardboard. Use a ruler to make sure your lines are straight. With the scissors, carefully cut out the rectangle you drew.

This piece will be the chassis, the main frame of the car.

2 At one end of your chassis, draw two dots, each 2 cm (¾ in) in from the end and from the side. Draw a line that passes through the dots.

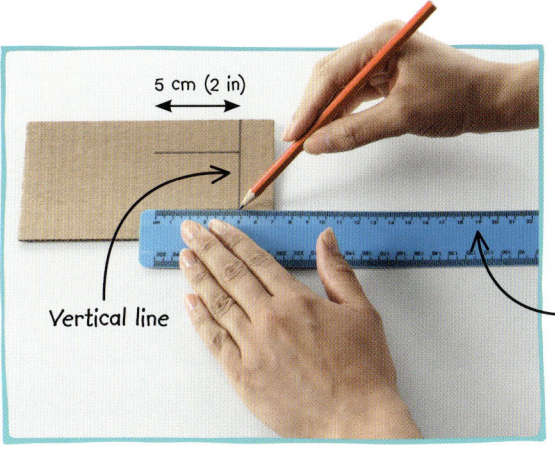

5 cm (2 in)

Vertical line

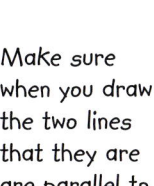

Make sure when you draw the two lines that they are are parallel to each other.

Ask an adult to help you if you find this step difficult.

3 Draw two lines, 5 cm (2 in) long at right angles from the vertical line you just drew, each one starting at one of the dots.

4 Using scissors, carefully cut along the middle of the vertical line, then down the two lines you just drew, to create a flap.

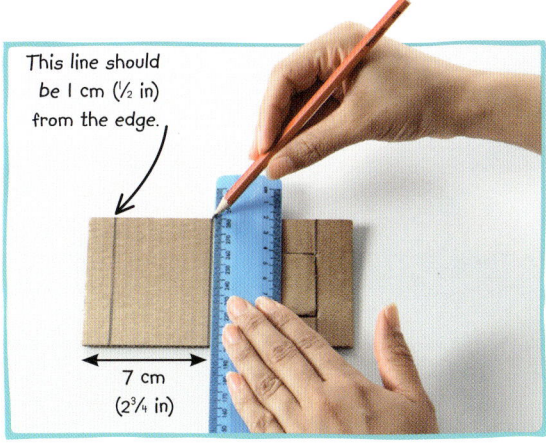

This line should be 1 cm (½ in) from the edge.

7 cm (2¾ in)

The dots should be 2 cm (¾ in) in from each end of this line.

5 Draw two more lines, parallel to the first one, about 1 cm (½ in) and 7 cm (2¾ in) from the other end of your chassis.

6 Make dots 2 cm (¾ in) in from each end of the line nearest the end, and draw a smooth curve from the dots to the ends of the other line. Cut along the curves.

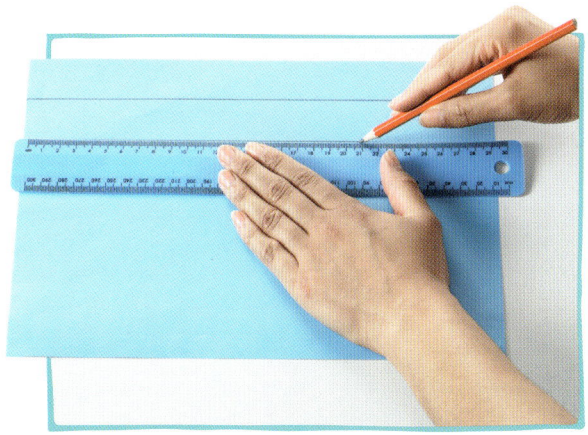

7 Paint the chassis. We've used green paint, but you can choose whatever colour you like.

8 On a piece of paper, draw two lines, 3 cm (1¼ in) and 6 cm (2½ in) in from one of the long sides of the paper.

Paper is a thin, versatile material made from mushed up wood fibres.

9 Cut along the two lines to make two long strips. These will be used to make the mainspring.

10 Use a small strip of double-sided tape to join the two pieces of paper together into one long piece.

The double-sided tape will allow you to seal the paper's top edge onto the tube.

Don't roll the paper too tightly, as your car's axle will need to turn inside it.

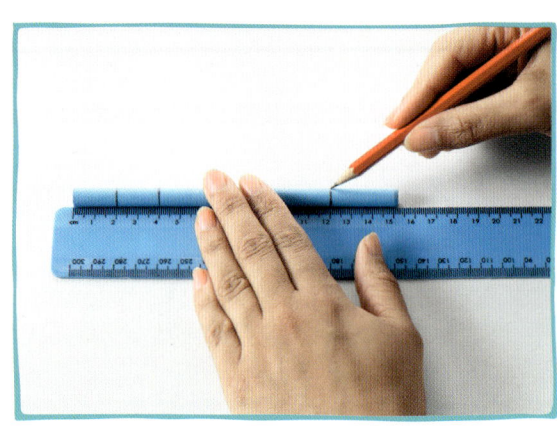

11 Take the rest of the sheet of paper and roll it lengthways around the garden stick to make a tube. Secure the tube with double-sided tape.

12 Draw lines on the tube at distances of 2, 4, and 12 cm (¾, 1½ and 5 in) from one end. These pieces will be your bearings.

Paper becomes very strong when it is rolled up.

13 Cut the tube along the lines you drew. You should end up with two pieces 2 cm (¾ in) long and one piece 8 cm (3 in) long. You don't need the rest of the paper tube, so try to recycle it.

14 Carefully cut two 11 cm (4½ in) lengths of garden stick. If you have trouble cutting it safely or neatly, ask an adult to help. These pieces will be your axles.

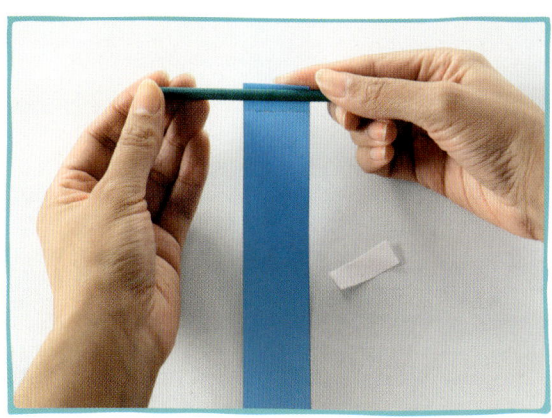

15 Tape one end of your long strip of paper to the middle of one of your garden stick axles. Coil the paper around the axle.

Once it is wound up, the car's mainspring will store potential energy.

Put the tape here.

16 Turn the chassis upside down and push the coiled mainspring through the flap of cardboard. Use double-sided tape to secure it.

Put one short paper tube onto this end of the axle.

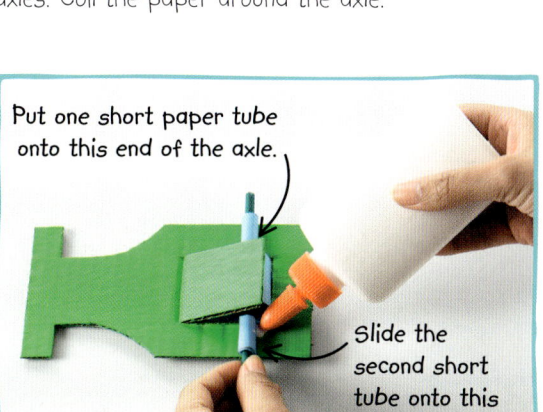

Slide the second short tube onto this end of the axle.

17 Turn the chassis over again, and slip one short paper tube over each end of the axle. Then glue them in place.

18 Slip the longer paper tube over the other garden stick axle and glue that in place near the other end of the chassis.

Leave the glue to dry thoroughly so it's really strong.

Be sure to protect the table and your fingers with adhesive putty.

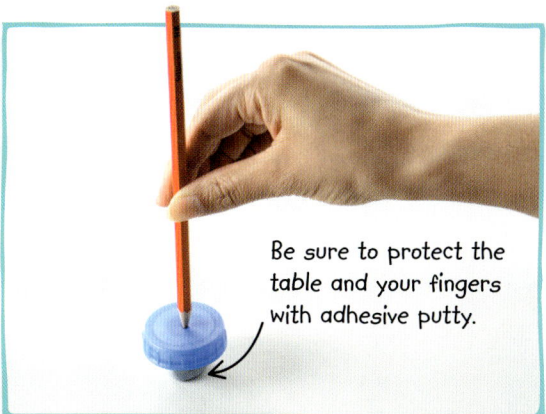

19 Use the pencil to make a small hole in the centre of each of the four bottle tops. Use adhesive putty to protect the table and your fingers.

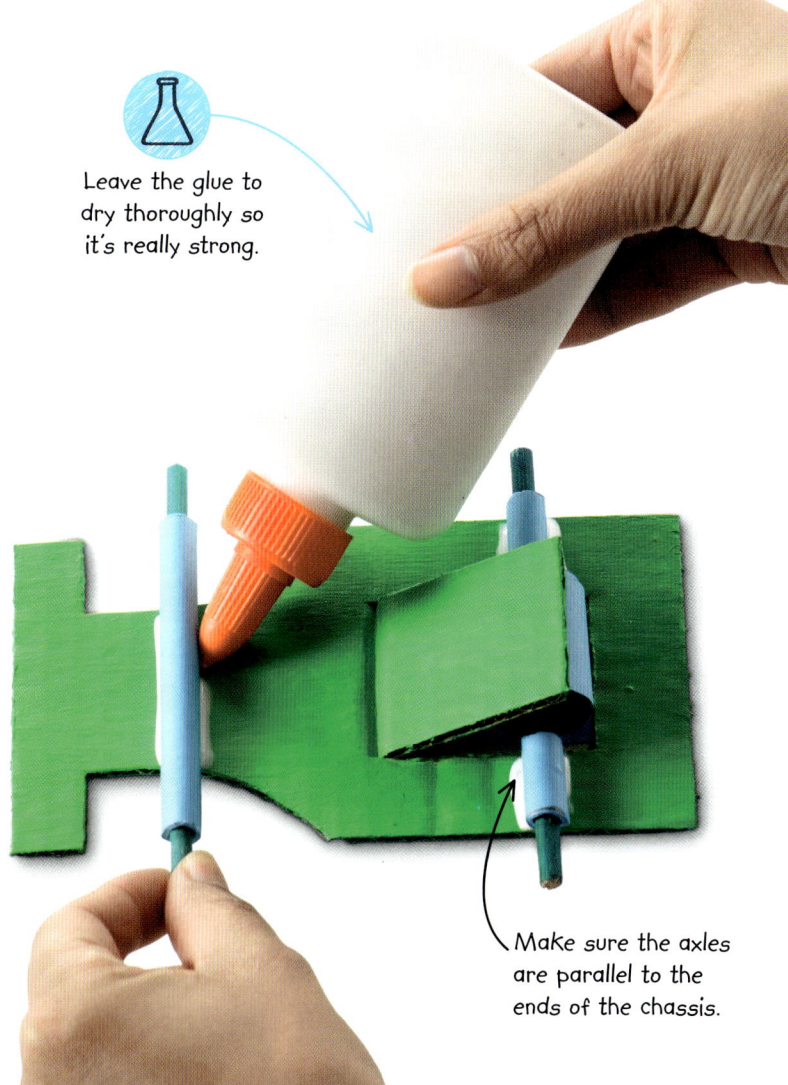

Make sure the axles are parallel to the ends of the chassis.

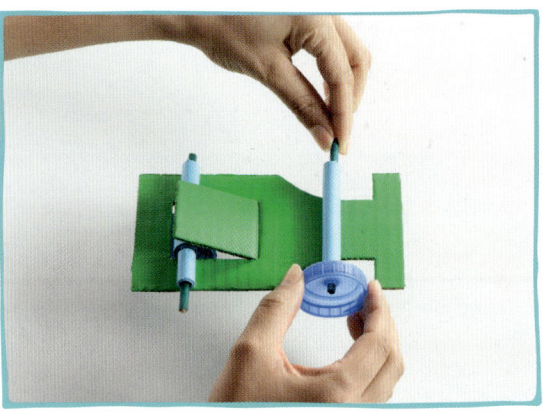

20 Push the *bottle tops* over the ends of the axle, to give your car wheels. If they are loose, secure them with adhesive putty or glue.

Energy can't be created or destroyed. It can only be transferred.

21 To make your car go, you have to wind up the mainspring. Do this by placing the car on the ground and pulling it *backwards.* Let go and watch it *speed off!*

You can work out your car's average speed by dividing the distance it travels by how long it takes.

The mainspring's energy is converted into kinetic energy, then lost as heat at the axles and ground, due to friction and air resistance.

HOW IT WORKS

Your car demonstrates potential and kinetic energy. Potential energy is stored energy, ready to make things happen. Kinetic energy is the energy objects have when they move. When you wind up the mainspring, you are storing potential energy, which will be used to make the car travel forwards. The faster an object moves and the more mass it has, the more kinetic energy it has. You can calculate the amount of energy a moving object has: multiply its mass (the amount of matter, or stuff it is made of) by its speed squared, then divide by two.

The mainspring is coiled up tightly.

1 As you pull the car backwards, the turning wheels coil the mainspring tightly, storing energy. When you let go, the spring uncoils and the potential energy becomes kinetic energy. The car moves forwards.

TEST AND TWEAK

Your wind-up car should zoom across the floor or table as the mainspring unwinds. Test it out on different surfaces and adapt its design to see if you can make your car go further and faster.

SANDPAPER WHEELS

Try wrapping sandpaper around the rear wheels to increase the amount of friction between them and the ground.

RUBBER BANDS

Putting rubber bands around the wheels gives the wheels extra traction, or grip, like the rubber tyres of a real car.

CARD MAINSPRING

A mainspring made of card should make your car go faster, as card stores more energy than paper. But it will release this energy faster, so your car won't travel as far.

REAL WORLD: TECHNOLOGY
ELECTRIC CARS

Most cars use the chemical potential energy stored in petrol to move, but not all. Electric cars have powerful batteries that store electrical potential energy. They can be recharged, like a smart phone.

REAL WORLD: MATHS
AIR RESISTANCE

Moving cars encounter a force called air resistance, which slows them down. Air resistance increases with speed. If you double the speed of the moving vehicle, the air resistance quadruples.

2 The spring continues uncoiling and the car keeps moving. Its kinetic energy is lost as heat. This happens through friction (at the axles and the ground) and air resistance.

The mainspring has completely uncoiled, and can provide no more energy for the car.

3 You can't feel the heat generated by friction and air resistance, as there isn't much kinetic energy in the first place. Once the kinetic energy is lost, the car stops.

BOTTLE RAFT

This activity could save your life! If you were stranded on a desert island, and you had some large empty barrels, you could make a raft to escape! It's a simple matter of balancing forces. The bowl of pebbles on the lollipop stick platform pushes the raft downwards into the water, but this force is balanced out by the buoyancy, or "upthrust", of the water pressing against the air-filled plastic bottles. Because these forces are equal, the raft floats!

The raft's platform is made of lollipop sticks, which are strong but light.

The bottles are filled with air, which makes them lighter than water.

This bowl of pebbles is acting as a load – a force that the raft's structure can withstand.

When the raft is placed in water, the water pushes upwards on the bottles, with a force called buoyancy, or "upthrust".

HOW TO BUILD A
BOTTLE RAFT

Empty plastic bottles float well in water, but to make an effective raft, you need to build a platform on which to support the load. It's a fairly simple project – the raft's platform is made of lollipop sticks glued together, and it is attached to the bottles with stretched rubber bands.

Time
30 minutes

Difficulty
Medium

STEM YOU WILL USE
• SCIENCE: Water pushes upwards on objects placed in it, using a force called "upthrust".
• ENGINEERING: Every structure can be made stronger by adding more support.

WHAT YOU NEED

Bowl of pebbles

Rubber bands

Glue

23 lollipop sticks

Scales

Two 500 ml (1 pint) bottles

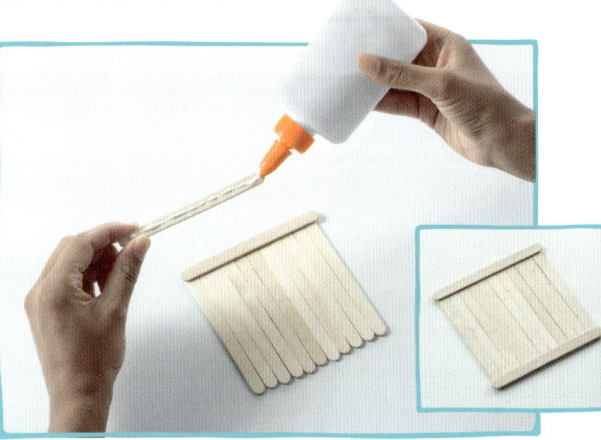

1 Lay eleven lollipop sticks side-by-side. Secure them together by adding glue to two other lollipop sticks and positioning them on either side of the platform.

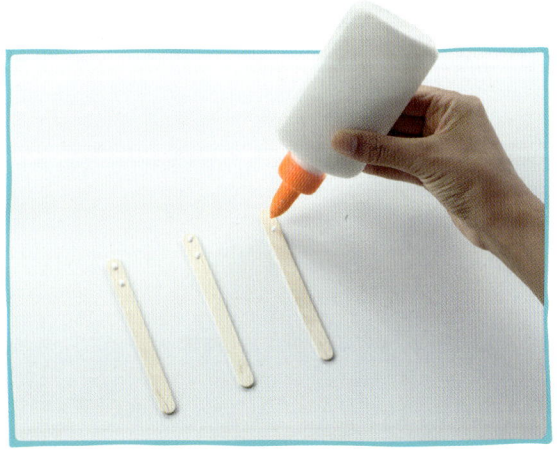

2 Take three lollipop sticks. Space them evenly so that they stretch the length of one lollipop stick. Put glue at the far end of each stick.

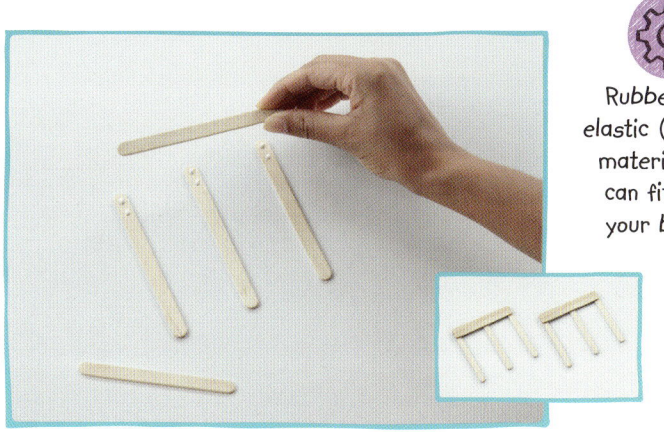

Rubber is an elastic (stretchy) material, so it can fit round your bottles.

3 Press two lollipop sticks on top of the dabs of glue to make an E-shape. Repeat steps 2 and 3 to make a second E-shape.

4 Once the glue on your two E-shapes has dried, slip two rubber bands over the ends of each one.

Before sticking the E-shapes to your platform, check that you have four rubber bands on each E-shape.

5 After you've placed the rubber bands onto both your E-shapes, turn the raft's platform over and glue the E-shapes onto it at both ends. Use lots of glue. Leave it to dry completely.

6 Stretch the rubber bands one-by-one enough to push the bottles through. Try to ensure that the rubber bands are evenly spaced.

The lids keep your raft airtight and watertight.

Rafts are usually built with light materials, such as wood, plastic, or foam.

8 Float your raft in the sink or bath, or even on a pond (make sure you have an adult with you). Gently place the bowl of pebbles on top of your raft's platform... can it take the load?

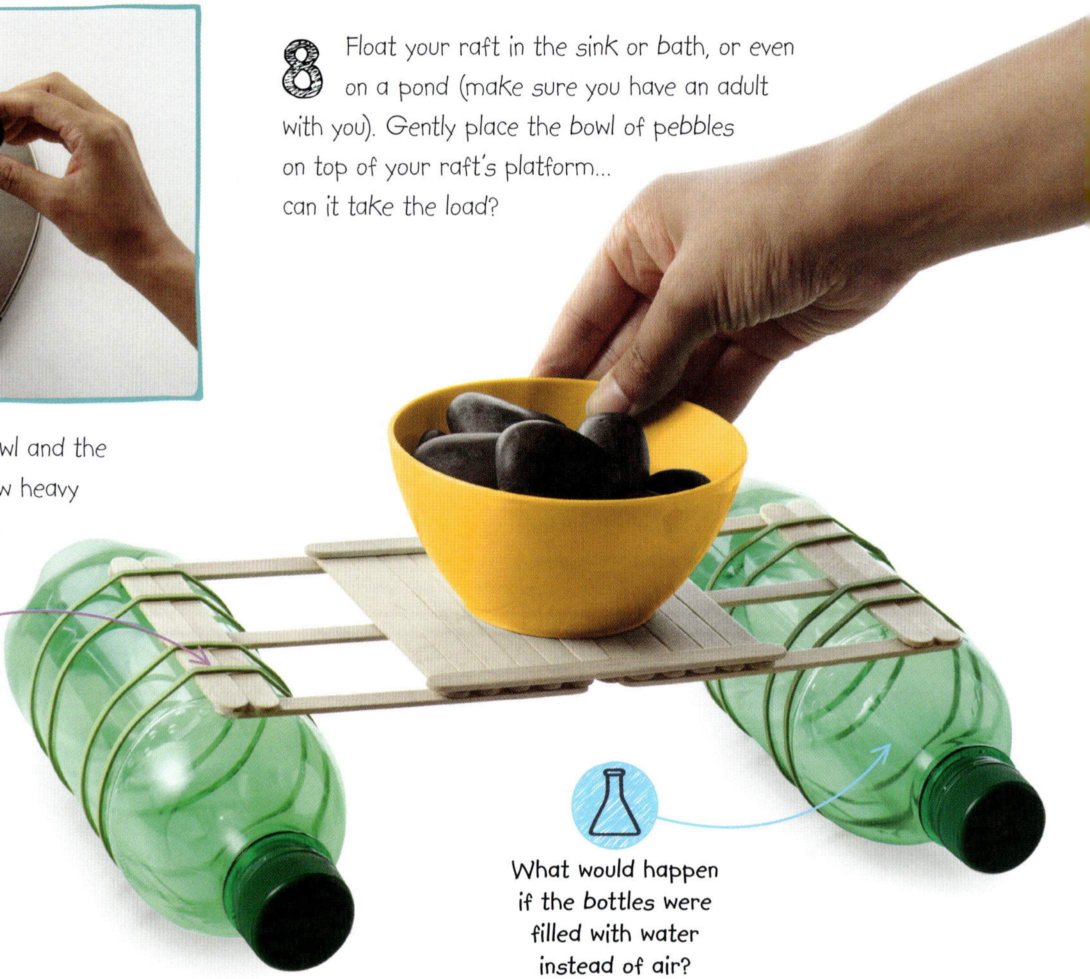

470g

7 Use scales to weigh the bowl and the pebbles, so you can see how heavy a load your raft is able to carry.

The strength of the join between the frame and platform might limit how heavy a load your raft can take – how can you make it stronger?

What would happen if the bottles were filled with water instead of air?

TAKE IT FURTHER

See how much weight your raft can support by experimenting with heavier loads. You could also adapt your raft to make a bridge or even a boat. To make a boat, add a sail to give it propulsion and a rudder underneath to help it steer a straight or curved path.

This large bowl of sand is heavier than the bowl of pebbles. What happens if you put it on your raft?

To support heavier loads, you could use bigger bottles or more bottles to make the raft more buoyant.

A pontoon is a bridge made by tying boats together. To turn your raft into a pontoon, simply add more platforms and bottles!

HOW IT WORKS

Whether or not an object floats depends on something called density. Density is how much mass (stuff) an object contains relative to its volume (the amount of space it takes up). When you place an object in water, the water pushes it upwards with a force called upthrust. If an object is more dense than water, the upthrust is too weak to support its weight, and the object sinks. That's why small, heavy things like coins and stones sink. Objects with low density, like your air-filled plastic bottles, are less dense than water, so the upthrust supports their weight and makes them float. Any object more dense than water will sink, and any object less dense will float.

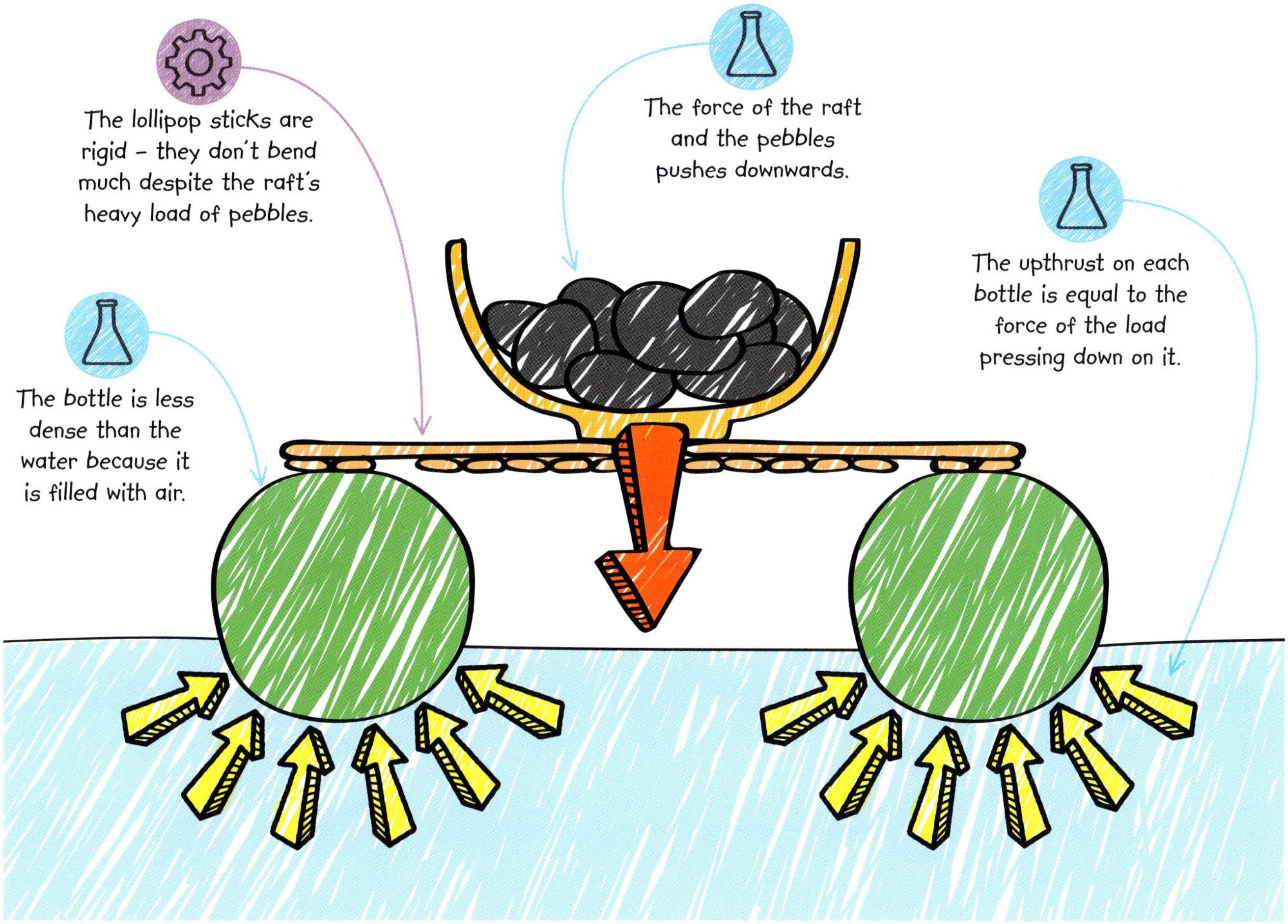

The lollipop sticks are rigid – they don't bend much despite the raft's heavy load of pebbles.

The force of the raft and the pebbles pushes downwards.

The upthrust on each bottle is equal to the force of the load pressing down on it.

The bottle is less dense than the water because it is filled with air.

REAL WORLD: ENGINEERING

SUBMARINES

Submarines can change their buoyancy – that's how they rise to the surface and dive deep. They have tanks that can be filled with water or air. At the surface, they take water into those tanks, increasing their density – so they sink. To rise up, air is pumped into the tanks, reducing their density and allowing them to float up to the surface.

FORCE OF GRAVITY
SAND PENDULUM

You can draw *beautiful* patterns with lines of sand by making a simple swinging device
called a pendulum. All you need is some sand, a plastic *bottle*, and a long piece of string.
This activity is great fun to watch but there's plenty of science to think about, too, like
how the force of gravity makes the pendulum swing back and forth.

As the swinging pendulum spirals inwards, it produces beautiful patterns.

We've used brightly coloured sand, but you could use ordinary sand, or salt.

HOW TO MAKE A
SAND PENDULUM

For this activity, you'll need plenty of space. We've used green-coloured sand, but ordinary sand is fine, too. Make sure your sand is perfectly dry, otherwise it won't flow freely. If you don't have sand, you can use salt instead.

Time
30 minutes

Difficulty
Medium

WHAT YOU NEED

String

Pencil

Scissors

Long ruler

Hole punch

Large sheets of dark paper

Plastic bottle

Sticky tape

Adhesive putty

Coloured sand, plain sand, or salt

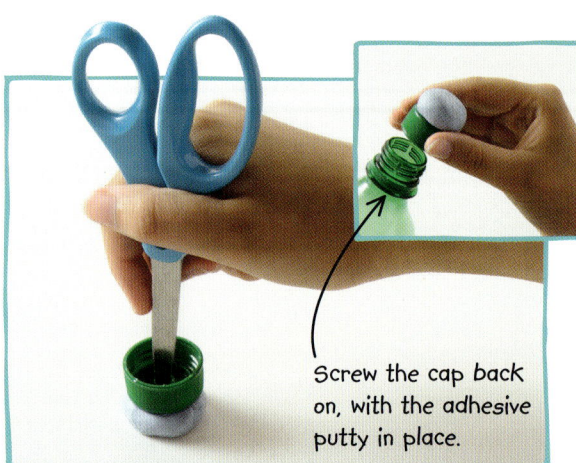

Screw the cap back on, with the adhesive putty in place.

1 Place the cap of the bottle upside down on a lump of adhesive putty. Use the scissors to make a hole about 3 mm (⅛ in) wide in the middle.

Ask an adult if you find this part tricky.

2 Using the scissors, cut off the bottom of the bottle. Try to keep a straight line.

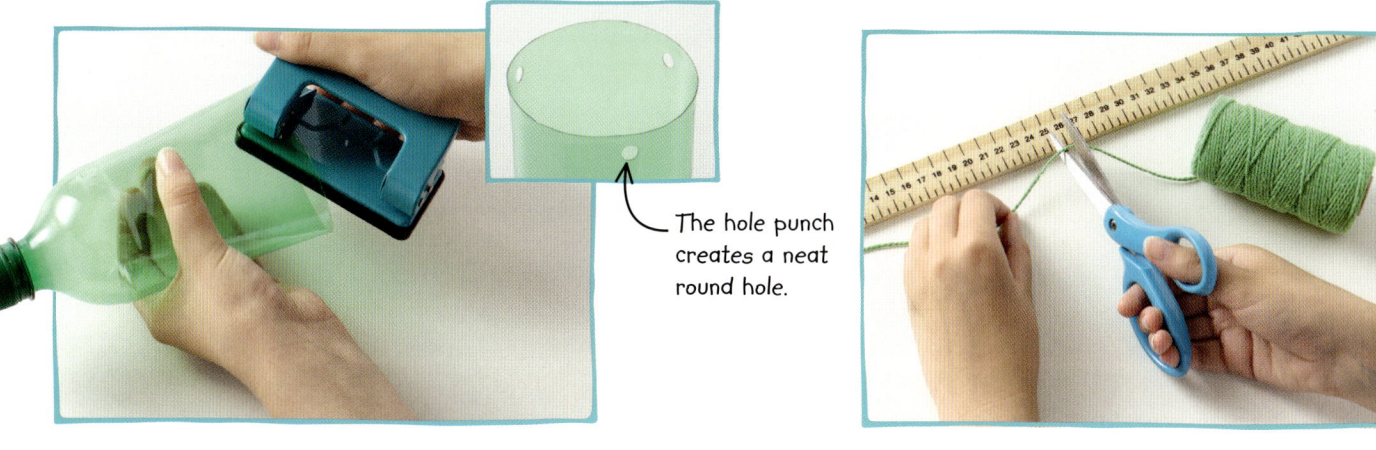

The hole punch creates a neat round hole.

3 Use the hole punch to make three evenly spaced holes in the plastic bottle, about 1 cm (½ in) away from the edge that you cut.

4 Measure and cut a piece of string about 25 cm (10 in) long.

Make sure the knot is secure.

5 Tie the string to two of the holes in the bottle to make a loop.

6 Cut a second length of string at least 2 m (7 ft) long. Tie one end of it to the third hole in the bottle.

Adjust the position of this knot to make the bottle hang straight.

7 Tie the long piece of string to the loop, taking care to keep the three lengths of string from each hole equal in length. This will help your bottle hang straight.

String is made from woven plant fibres.

8 Ask an adult to help you suspend the pendulum from a high point (such as the branch of a tree or a hook on a ceiling) so the bottle cap is 4–5 cm (2 in) above the ground. Pour sand or salt into the bottle.

9 Use the sticky tape to join a few sheets of the dark paper. This will make one large piece to catch the sand that falls from the bottle.

10 Remove the adhesive putty and give the bottle a gentle sideways push to make it swing in a circle. Once the bottle is empty, fold up the paper and tip the sand back into the bottle. You can then try the experiment again.

The pendulum slowly loses energy due to friction between the string and the point where it is tied, and air resistance between the bottle and the air.

The bottle moves in oval shapes called ellipses.

The ellipses get smaller as the pendulum loses energy.

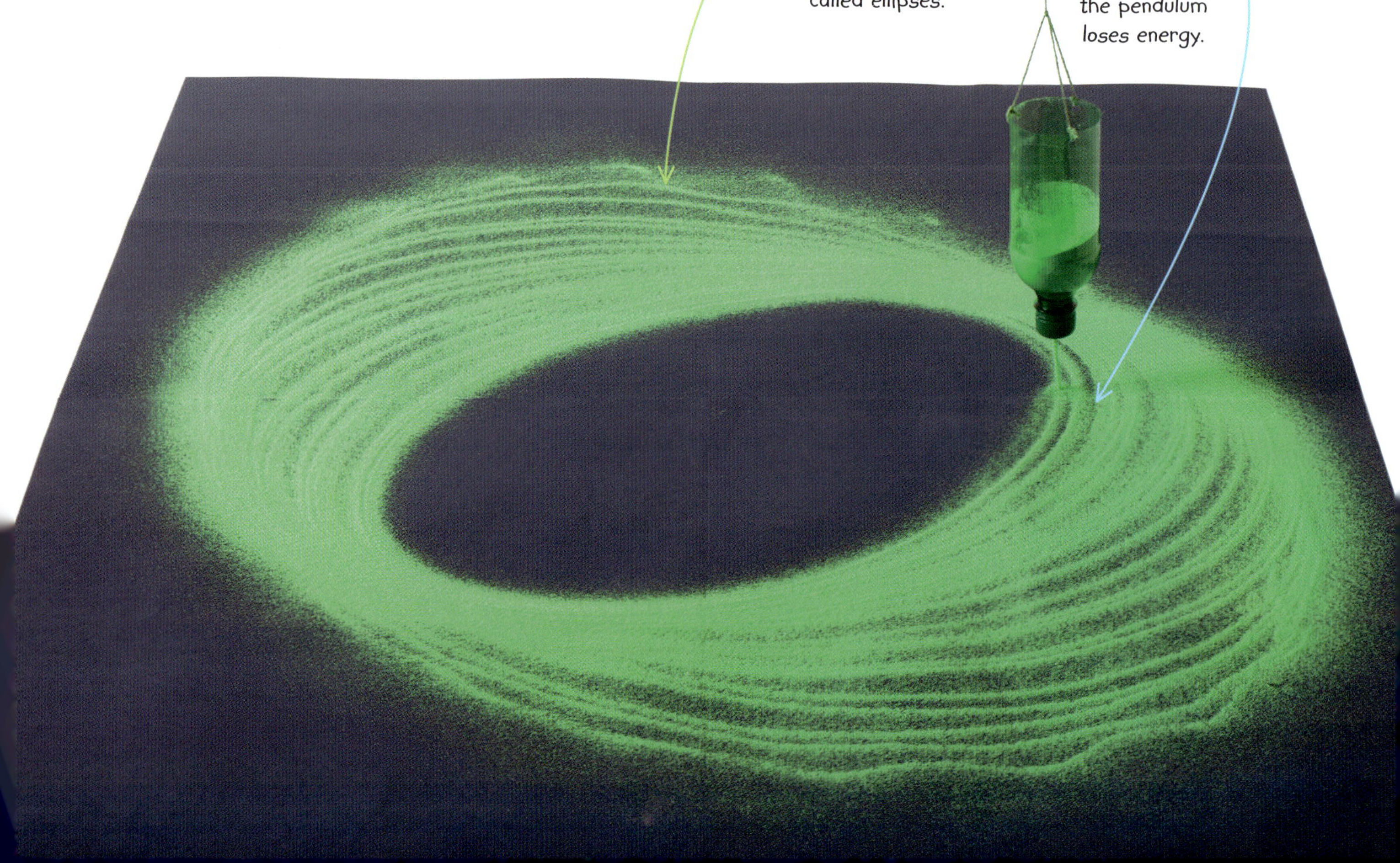

TEST AND TWEAK

In the 1580s, the Italian scientist Galileo discovered that a pendulum swings back and forth in a straight line for a precise time, or period, that depends on its length – a discovery that eventually led to the invention of pendulum clocks. Try changing the length of your pendulum to see how it affects the time it takes to swing back and forth in a straight line. You can also make the pendulum's elliptical movements more complex by making the string Y-shaped. This gives the pendulum a short period in one direction and a long period in another, resulting in weird and wonderful sand patterns known as Lissajous curves. If you raise or lower the meeting point between the Y's arms, the Lissajous curves will change.

Changing the position of this knot results in different sand patterns.

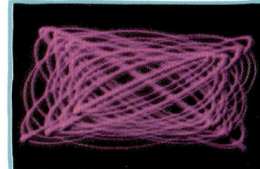

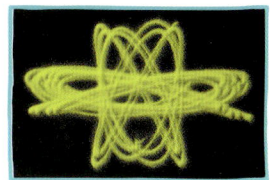

HOW IT WORKS

If you simply pulled your pendulum away from its resting point and let go, it would swing back and forth in a straight line until it ran out of energy. Because you pushed it sideways, it swung along a curving path – an ellipse – continually changing direction. A moving object only changes direction when a force acts on it. In this case, the force of gravity is pulling the bottle back to the middle, but its sideways motion and the pull of the string stop it returning directly. The pendulum loses energy due to friction. As a result, it slowly spirals inwards, the sand tracing out a beautiful record of its path.

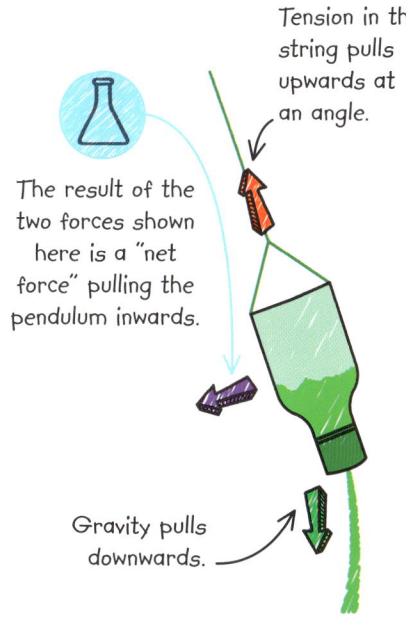

Tension in the string pulls upwards at an angle.

The result of the two forces shown here is a "net force" pulling the pendulum inwards.

Gravity pulls downwards.

FROM THE SIDE

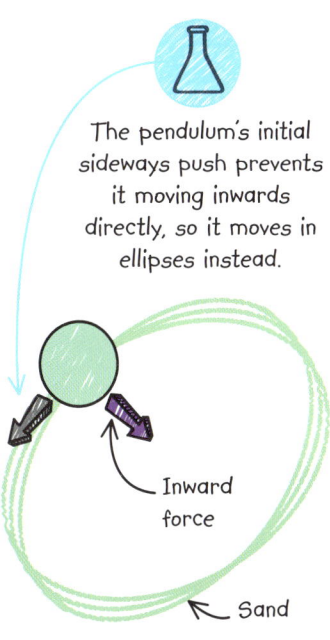

The pendulum's initial sideways push prevents it moving inwards directly, so it moves in ellipses instead.

Inward force

Sand

FROM THE TOP

REAL WORLD: SCIENCE
ORBITING OBJECTS

The artificial satellites orbiting our planet travel in ellipses or circles, just like your sand pendulum. There is just one force acting on all these orbiting objects: gravity. It pulls the satellites inwards, causing them to curve continually around Earth rather than flying off into space.

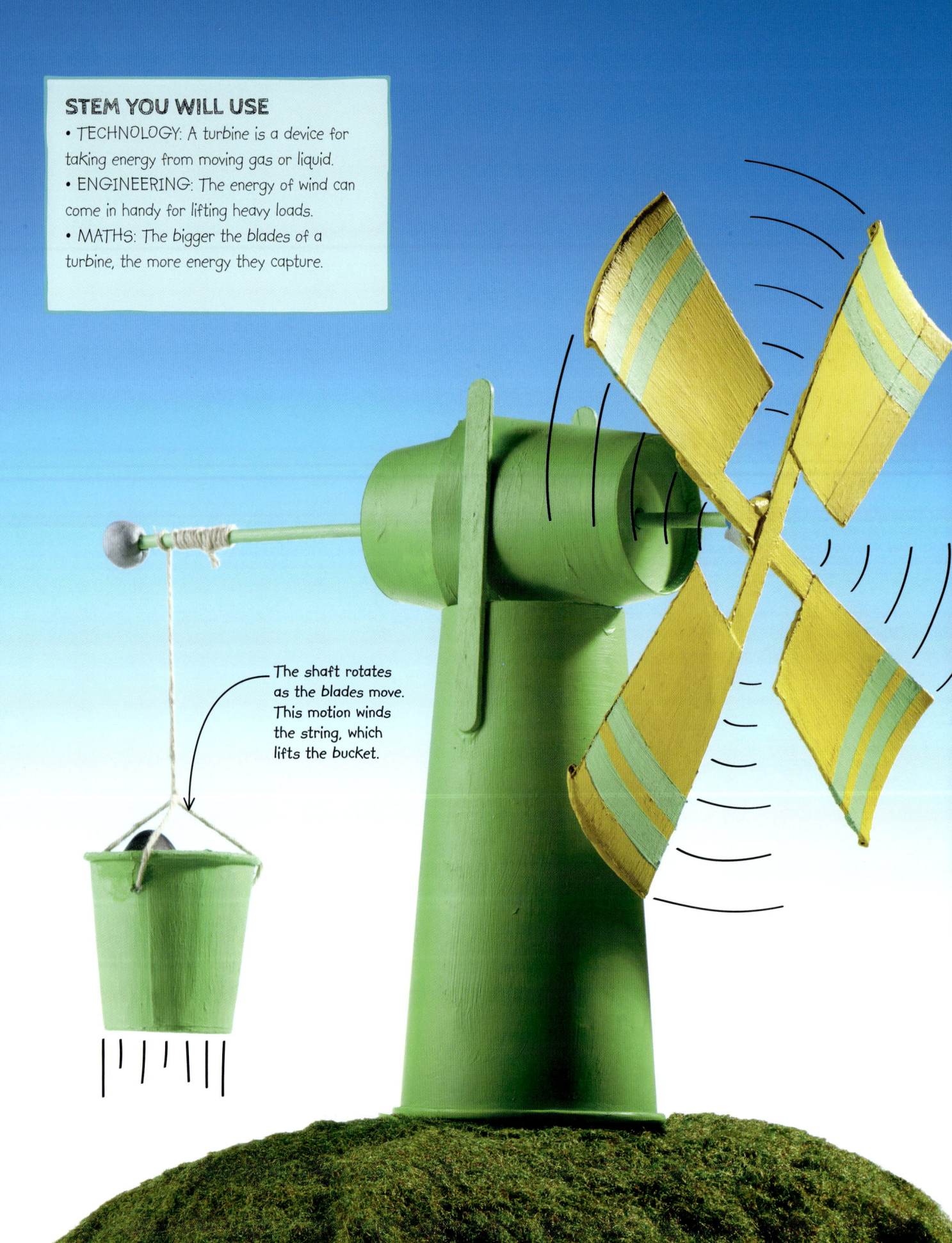

STEM YOU WILL USE
• TECHNOLOGY: A turbine is a device for taking energy from moving gas or liquid.
• ENGINEERING: The energy of wind can come in handy for lifting heavy loads.
• MATHS: The bigger the blades of a turbine, the more energy they capture.

The shaft rotates as the blades move. This motion winds the string, which lifts the bucket.

WIND TURBINE

Have you ever seen huge wind turbines spinning slowly around? The blades are being pushed round by the energy of the wind. Inside each tower is an electrical generator, which converts wind energy into electrical energy to power homes, offices, factories, and schools. You can explore the engineering challenge of extracting energy from the wind by building your own wind turbine, using paper cups to make the blades.

The curved blades deflect the wind. This makes the blades move in the opposite direction.

HOW TO MAKE A
WIND TURBINE

Perhaps the most important feature of a wind turbine is the fact that the blades are at an angle, and so deflect the wind. This turbine's blades are made from paper cups, which are naturally curved, so they deflect the wind and work well. Take time to make your turbine, waiting for the glue to set where necessary.

Time
45 minutes

Difficulty
Medium

WHAT YOU NEED

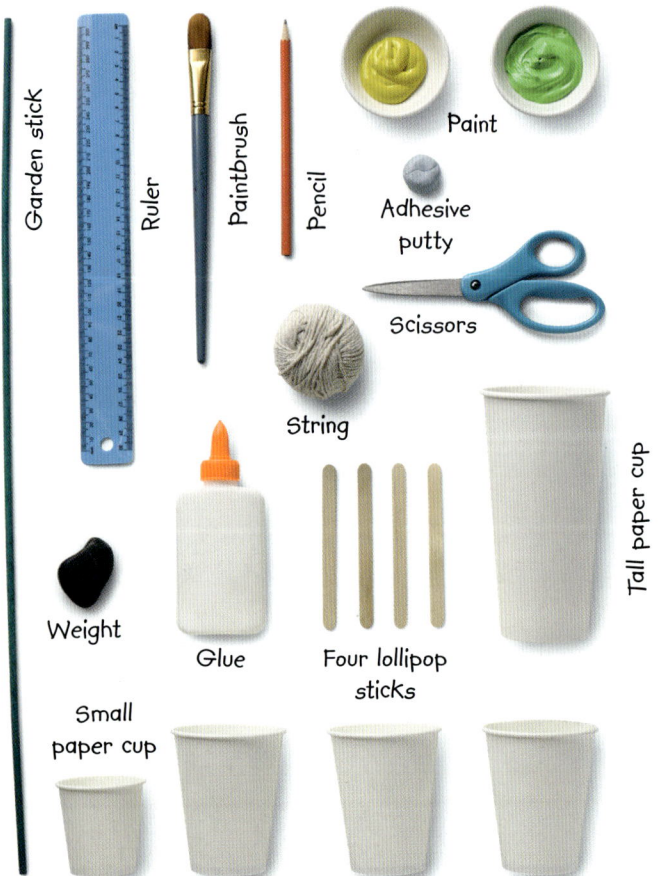

Garden stick

Ruler

Paintbrush

Pencil

Paint

Adhesive putty

Scissors

String

Weight

Glue

Four lollipop sticks

Tall paper cup

Small paper cup

Three medium paper cups

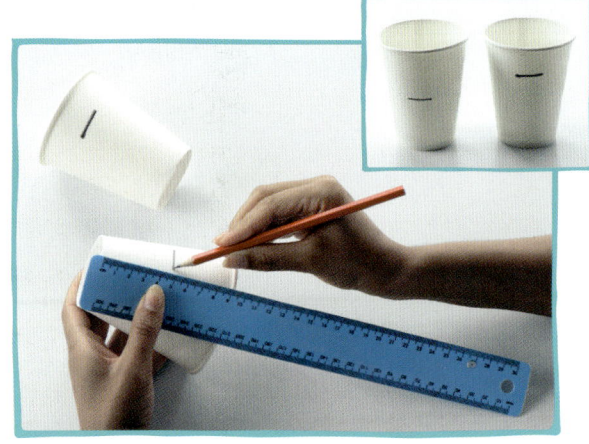

1 Take two medium cups and draw a line on the side of one 7 cm (2¾ in) from the bottom. Draw a line 5 cm (2 in) from the bottom of the other cup.

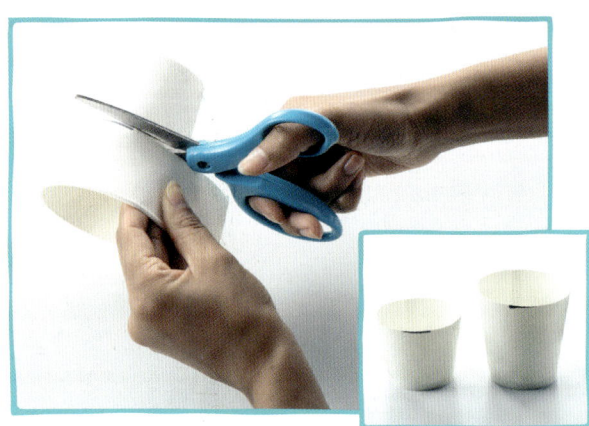

2 With a pair of scissors, carefully cut around the lines and remove the top parts of both cups. Discard the tops – recycle them if possible.

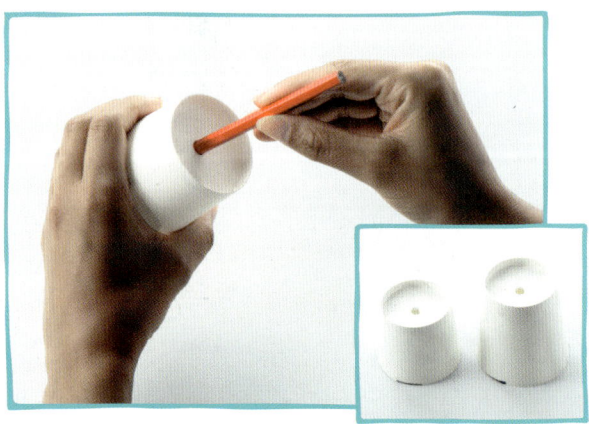

3 Using the sharp point of the pencil, pierce a hole at the centre of the base of each medium cup. Take care not to pierce yourself!

The shorter cup is upside down.

4 Insert the smaller of the two shortened cups into the larger one. Then squeeze glue into the joint to fix them together and wait for the glue to dry.

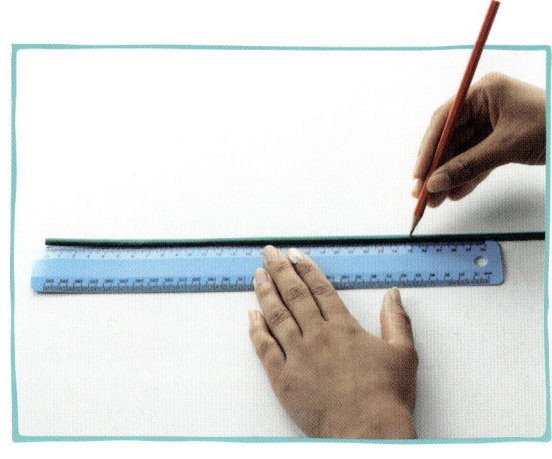

5 Make a pencil mark 25 cm (10 in) from one end of the garden stick.

Use scissors to score the stick.

6 Cut the garden stick at the pencil mark. Score the stick with scissors first, then *bend* it to snap it. Ask an adult if you find this tricky.

This stick performs the job of the shaft in a real wind turbine, which helps generate electricity.

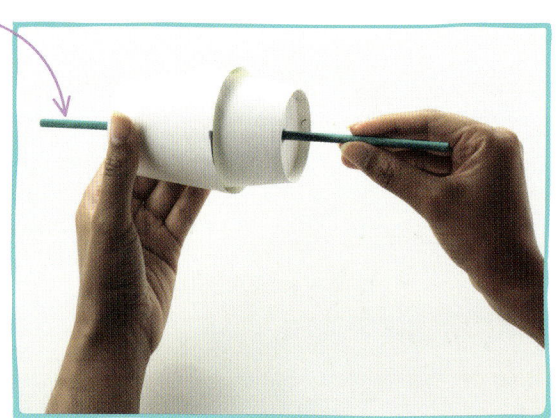

7 Slide the stick through the holes in the bases of the two joined cups.

8 Place the tall, uncut cup upside down and glue a lollipop stick to either side of the base, making sure each one reaches the same distance above the cup.

9 When the glue has dried, spread glue on the inside surfaces of the two lollipop sticks, near their ends.

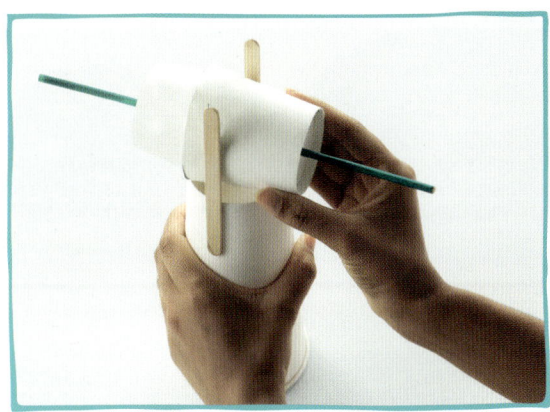

10 Place the joined cups with the stick through the centre between the lollipop sticks. Hold the cups in place while the glue dries.

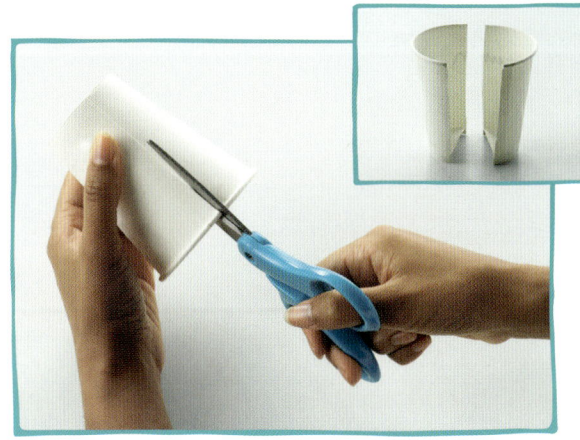

11 To make the turbine blades, take your remaining medium-sized cup and carefully cut it in half down the side with a pair of scissors.

12 Cut each half in half again so you are left with four equal pieces. Cut the base of each quarter off and recycle these pieces.

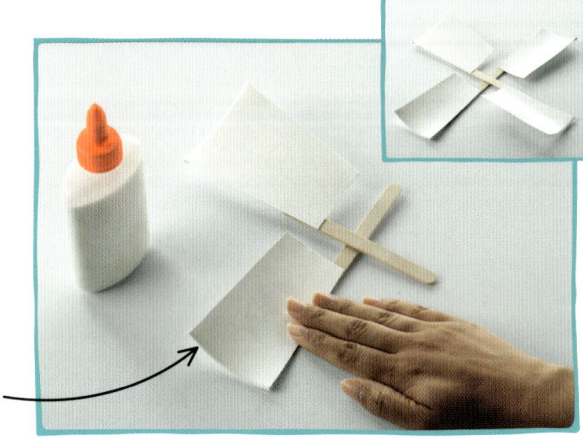

Make sure the blades all face the same way.

13 Place glue at the centre of a lollipop stick and stick it to another to form a cross. Glue the edges of your blades to the lollipop sticks.

On a real wind turbine, the blades on their shaft are able to swing around to face the wind.

14 Stick a piece of adhesive putty to the centre of the cross. The adhesive putty will secure the blades to the shaft.

15 To attach the blades to the wind turbine, fix the adhesive putty to the end of the stick in the top of the turbine.

Hold some adhesive putty on the inside of the cup to stop you hurting yourself.

16 Take the small cup and make three equally spaced small holes around the top using a sharp pencil. This will be your load-lifting bucket.

The string will act as the bucket's handle.

17 To connect the bucket to the wind turbine, cut a 12 cm (5 in) piece of string. Thread the string through two of the holes in the bucket and tie a knot at either end to secure it in place.

The wind turbine's blades are curved, helping them to deflect the wind.

The garden stick acts as a pulley, drawing up the string as it turns.

18 Measure and cut a 40 cm (16 in) piece of string. Thread one end of the string through the third hole in the bucket, then tie it to the middle of the short piece of string.

To neaten up the end of the stick, cover it with adhesive putty.

19 Tie the free end of the long string to the garden stick. If you want to be sure it won't slip, secure it with a small piece of tape.

20 Now paint and decorate your wind turbine in your favourite colours and patterns.

The curved blades of this wind turbine transfer some of the Kinetic (movement) energy in the wind into rotary (turning) motion in the blades.

Stick some modelling clay inside the base to act as a weight if the wind turbine Keeps falling over.

21 Now you can try it out! Put weights in the bucket and see how quickly it rises when you expose the turbine to wind. If there's no wind, you could use a fan or a hairdryer. What happens to the bucket when the wind stops – does it fall back down or does friction hold it in place?

Put different weights in the bucket to see how much your wind turbine can lift.

TAKE IT FURTHER

If you have a fan with different speed settings, investigate how quickly the windmill lifts the bucket as the wind speed increases. Try making different kinds of turbine blade to see which turns fastest. To test your designs fairly, use a fan and make sure you have it on the same speed setting each time. Can you make your turbine lift heavier weights?

LARGER BLADES

To make larger blades for your model simply use a larger cup.

MORE BLADES

Make a cross out of three lollipop sticks and cut up two cups to make more blades.

HOW IT WORKS

Wind is simply moving air. It is caused by uneven heating of Earth's surface by the Sun. In hot places, the warm air rises, causing cooler air to be drawn into the space left behind and so creating wind. For instance, land heats up under the Sun more quickly than the sea, so on sunny mornings a breeze often blows from sea to land. Wind turbines harness the kinetic energy of the wind to cause a generator inside the turbine to make electricity.

High in the sky, air cools down again and begins to sink.

Air over the land warms up and rises.

Cool air flows into the space left by the rising air, creating wind.

Land heats up faster than the sea.

REAL WORLD: TECHNOLOGY GENERATING POWER

Wind turbines use the kinetic energy in wind to generate power. Wind causes the turbine's blades to turn, which causes a generator in the main shaft of the turbine to spin. The generator produces electrical energy, which can be used to power things. Wind turbines produce the most energy in windy places, such as hilltops and on the coast.

LEVITATING BALL

Levitation is when something is lifted into the air with no visible means of support. Stage magicians pretend they are making things levitate, claiming they are using mysterious magical powers. But it isn't magic – usually a string is holding up the object. You can make a ping-pong ball levitate with no strings attached and without touching it. It looks like magic but it's science! The ball is held up by forces working against each other.

A fast jet of air comes from the straw when you blow into the wide tube.

The jet of air supports the ball, even when the ball is not directly over the end of the straw.

STEM YOU WILL USE
• SCIENCE: Air moves from high pressure to low pressure.
• TECHNOLOGY: Pipes and tubes are used to carry fluids – liquids or gases – from one place to another.

HOW TO MAKE A
LEVITATING BALL

To make the ping-pong ball levitate, you need to create a jet of air. In this activity, you do that by blowing into a cardboard tube attached to a piece of drinking straw. For the strongest jet of air, you need to make sure there are no leaks in the tube or straw.

Time
20 minutes

Difficulty
Easy

WHAT YOU NEED

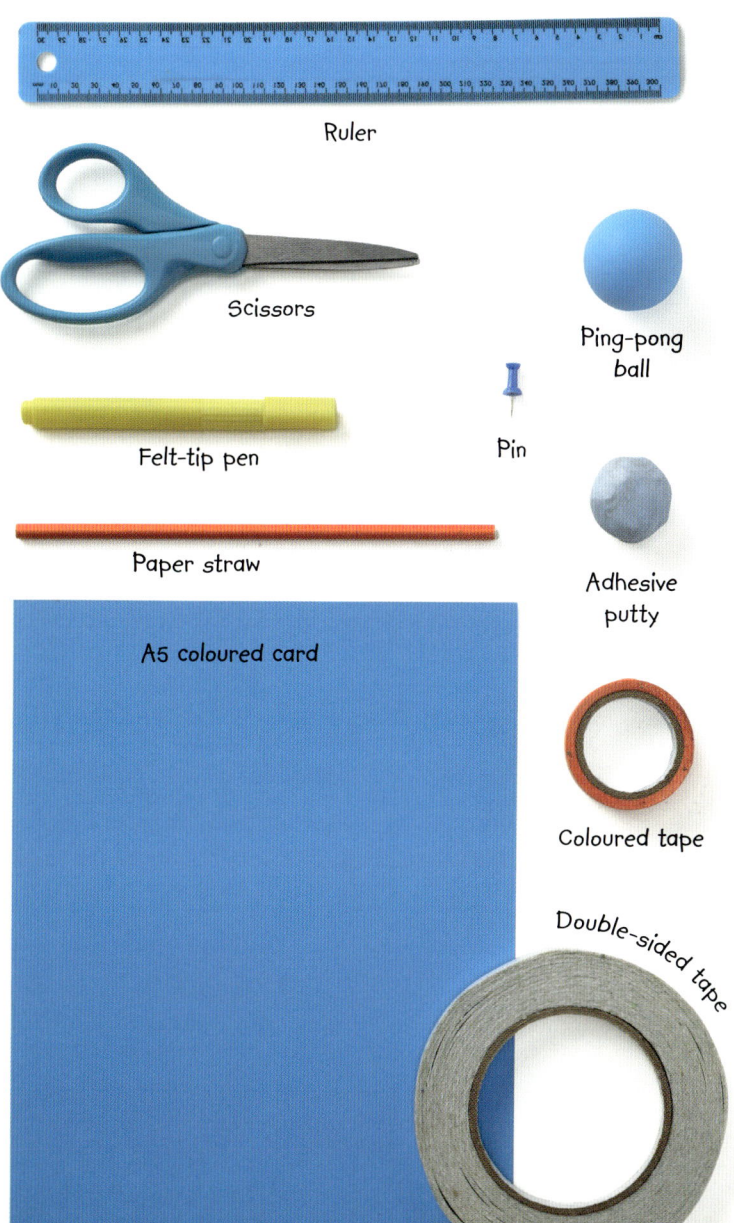

Ruler

Scissors

Felt-tip pen

Pin

Ping-pong ball

Adhesive putty

Paper straw

A5 coloured card

Coloured tape

Double-sided tape

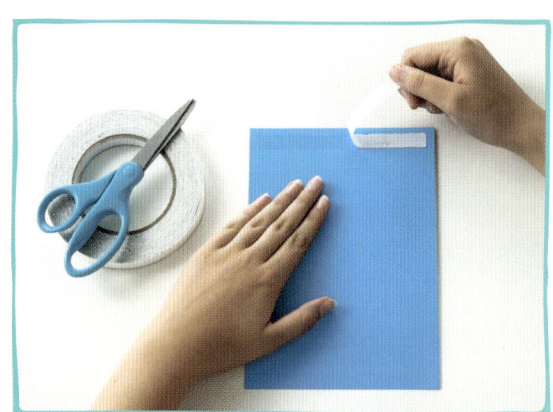

1 Place a piece of double-sided tape across one of the short edges of the A5 piece of coloured card. Remove the tape's protective strip.

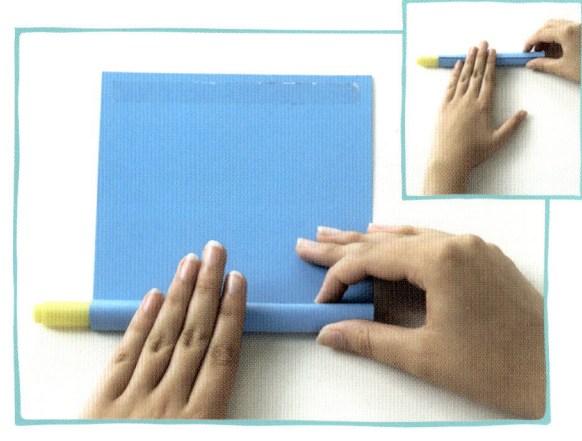

2 Starting at the short edge without the tape on it, roll the card around the felt-tip pen to create a tube. Press the sticky tape down firmly once you reach the end. Remove the pen.

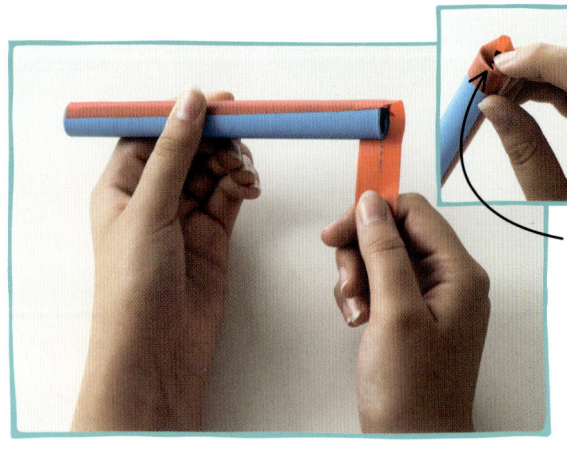

Fold down the
tape to form
a neat edge, but
ensure you keep
this end open –
don't cover it.

3 Stick down the edge of the card with some
coloured tape. Then wrap more coloured
tape around one end of the tube.

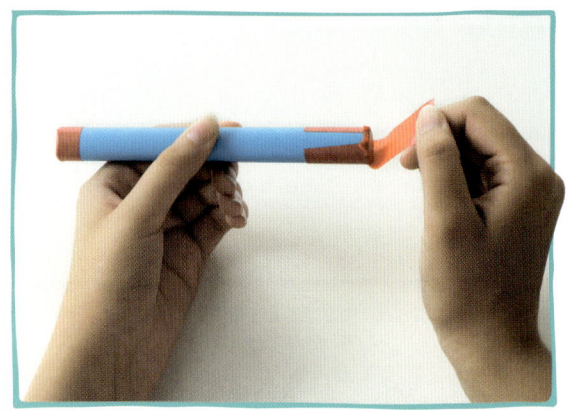

4 Now close off the other end completely with
some more coloured tape. This will ensure that
no air can escape when you blow into the other end.

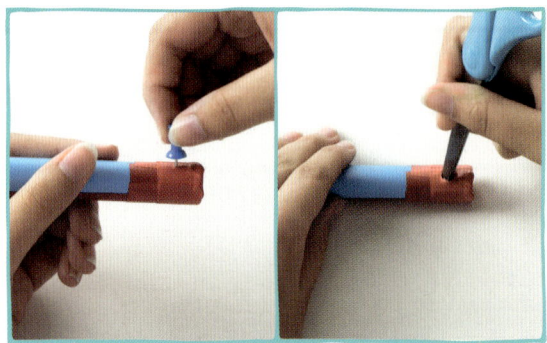

5 Use the pin to make a small hole through the
coloured tape and the card near the closed
end of the tube. Gently make the hole a bit bigger
with the point of the scissors.

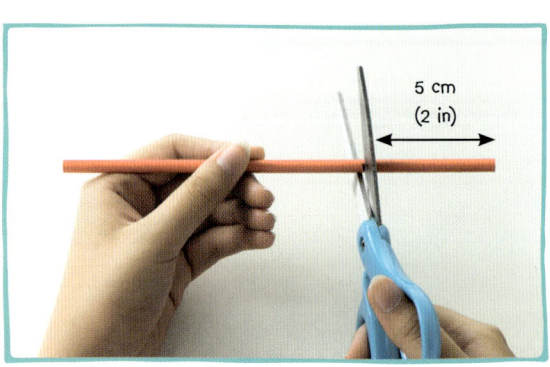

5 cm
(2 in)

6 Cut a piece of drinking straw that is about
5 cm (2 in) long. Recycle the rest if you can.

Carefully cut
about 1 cm (½ in)
into the straw.

7 Carefully make two cuts, just over 1 cm
(about ½ in) long, on either side of one end
of the straw. Fold up one side in between
the two cuts to make a flap.

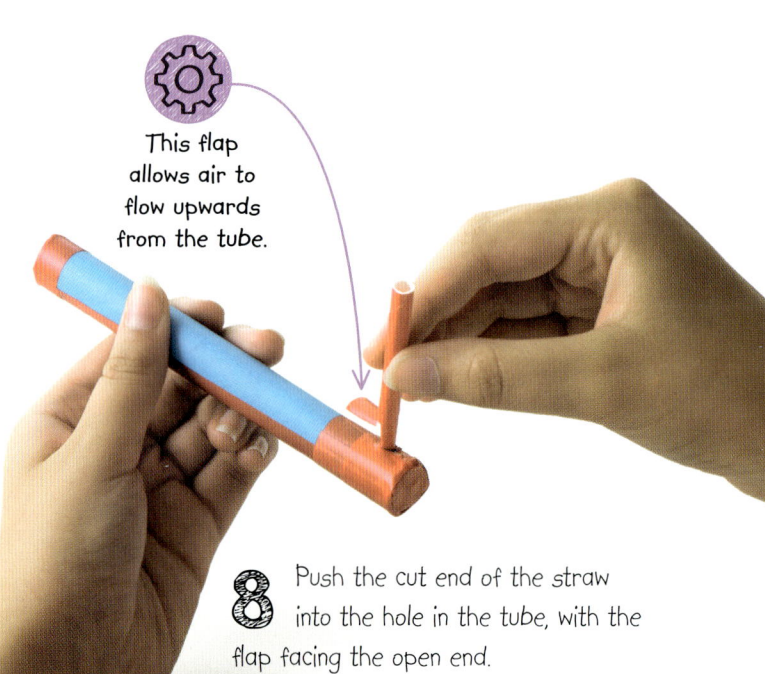

This flap
allows air to
flow upwards
from the tube.

8 Push the cut end of the straw
into the hole in the tube, with the
flap facing the open end.

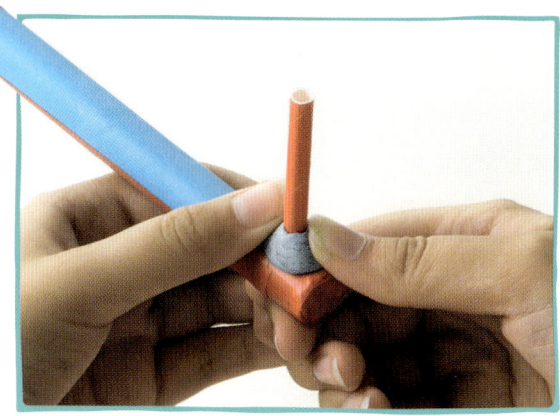

10 Your levitation tube is complete. Hold the ping-pong ball above the straw, blow into the open end of the card tube, and let go of the ball. Can you make the ball levitate?

9 Use coloured tape to secure the flap to the tube. Then wrap adhesive putty around the base of the straw, where it meets the card tube. Don't press so hard that you close off the airway, but do try to block any leaks.

The air speeds up as it moves from the wide tube to the narrow straw.

HOW IT WORKS

When you blow through the tube, the stream of air pushes the ball, lifting it. Even if you tilt the tube slightly, the ball doesn't drop. This is because air flowing past a smoothly curved object (such as your ball) will curve to flow over its surface and so bend sideways as it leaves. The ball has effectively pushed the airflow sideways, and this results in a "reaction force" that pushes back against the ball. The reaction force stops the ball from falling.

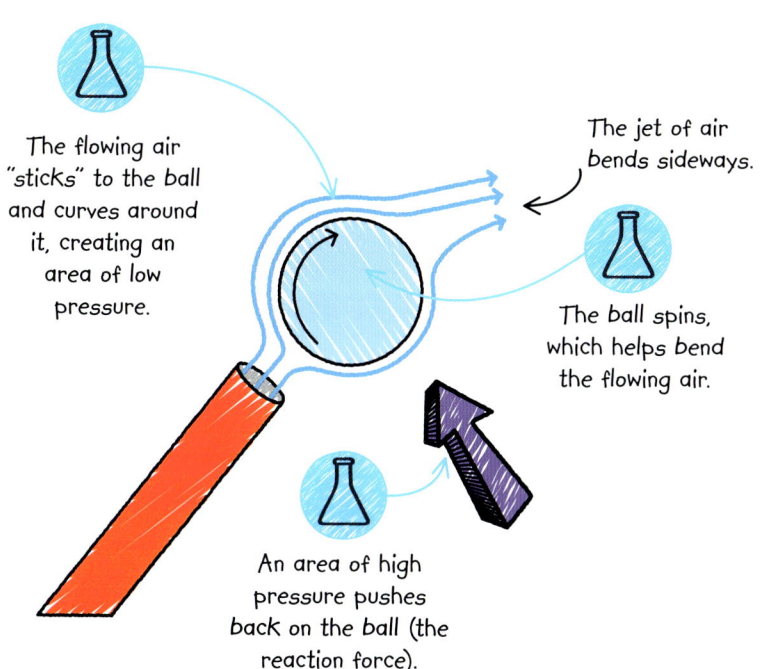

The flowing air "sticks" to the ball and curves around it, creating an area of low pressure.

The jet of air bends sideways.

The ball spins, which helps bend the flowing air.

An area of high pressure pushes back on the ball (the reaction force).

REAL WORLD: SCIENCE
THE POWER OF SAILS

Sails also work by changing the direction of airflow. When wind hits a sail, it follows the curve of the sail and so is deflected in a different direction. This causes a reaction force that pushes the craft the opposite way. By adjusting the angle of the sail, you can sail in almost any direction – even into the wind!

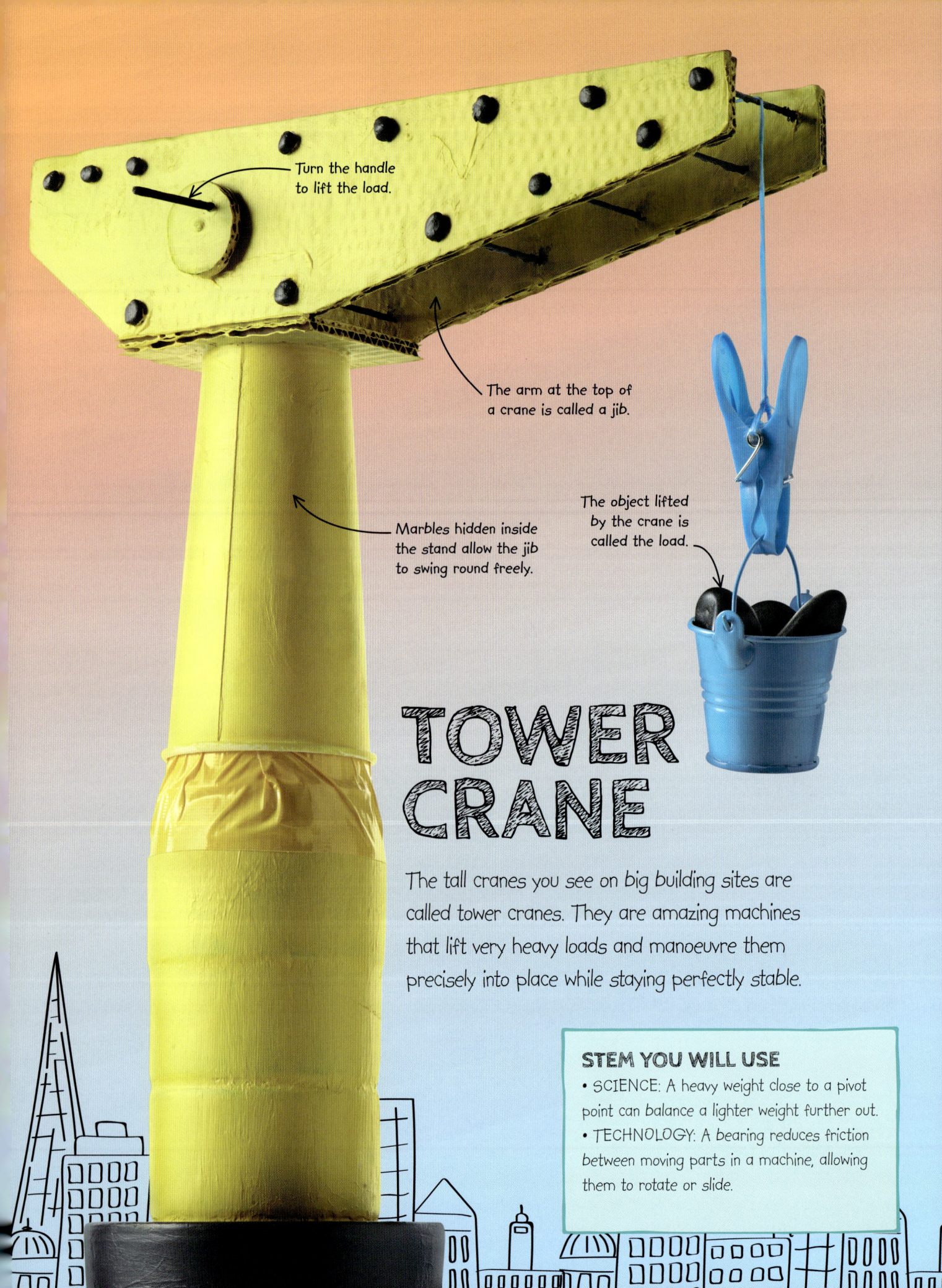

Turn the handle to lift the load.

The arm at the top of a crane is called a jib.

Marbles hidden inside the stand allow the jib to swing round freely.

The object lifted by the crane is called the load.

TOWER CRANE

The tall cranes you *see* on big building sites are called tower cranes. They are amazing machines that lift very heavy loads and manoeuvre them precisely into place while staying perfectly stable.

STEM YOU WILL USE
• SCIENCE: A heavy weight close to a pivot point can balance a lighter weight further out.
• TECHNOLOGY: A bearing reduces friction between moving parts in a machine, allowing them to rotate or slide.

HOW TO BUILD A
TOWER CRANE

You'll need patience for this build as there are lots of steps. The trickiest part is the jib – the horizontal arm on top of the crane. It's made of two pieces of cardboard held together by toothpicks. The crane's stand is made from heavier materials that keep the whole structure stable.

Time
1 hour and 30 minutes

Warning
This activity uses small marbles. Don't put them in your mouth.

Difficulty
Hard

WHAT YOU NEED

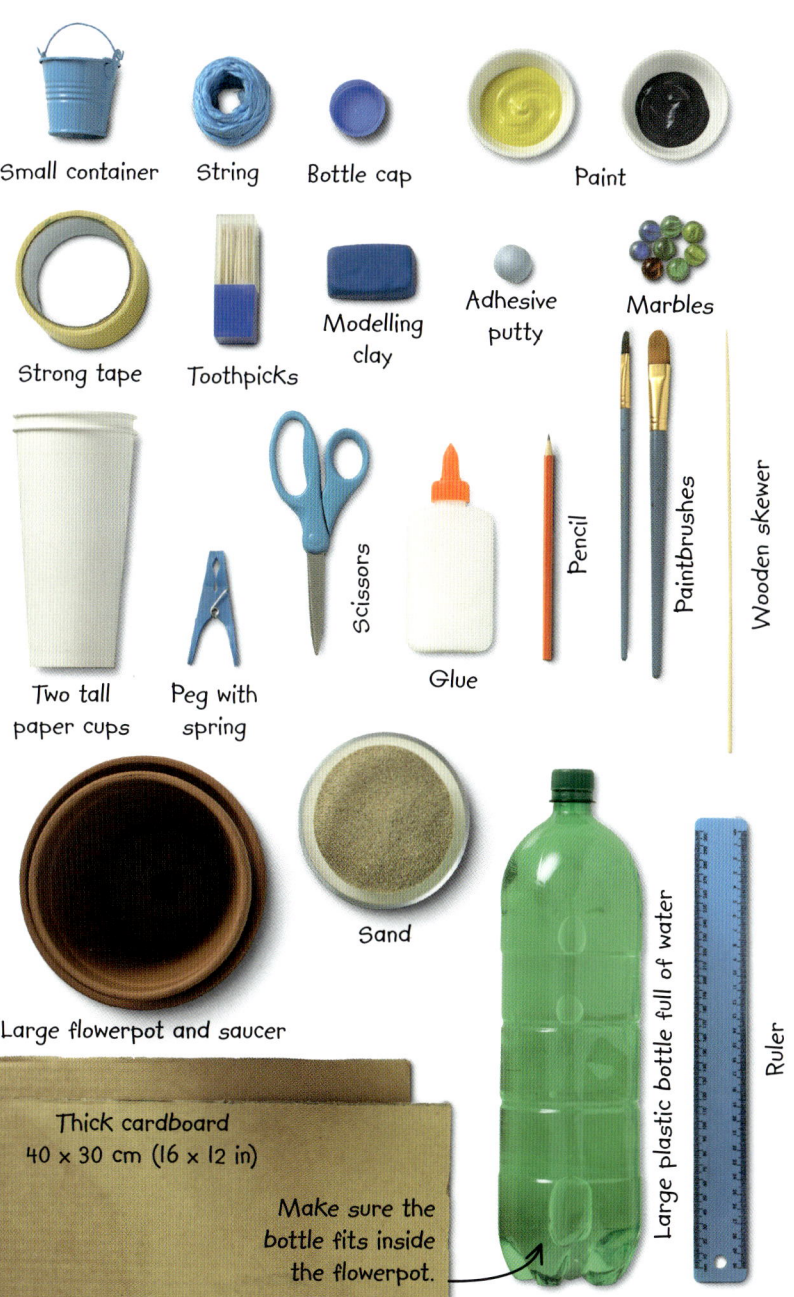

Small container

String

Bottle cap

Paint

Strong tape

Toothpicks

Modelling clay

Adhesive putty

Marbles

Two tall paper cups

Peg with spring

Scissors

Glue

Pencil

Paintbrushes

Wooden skewer

Large flowerpot and saucer

Sand

Large plastic bottle full of water

Ruler

Thick cardboard
40 x 30 cm (16 x 12 in)

Make sure the bottle fits inside the flowerpot.

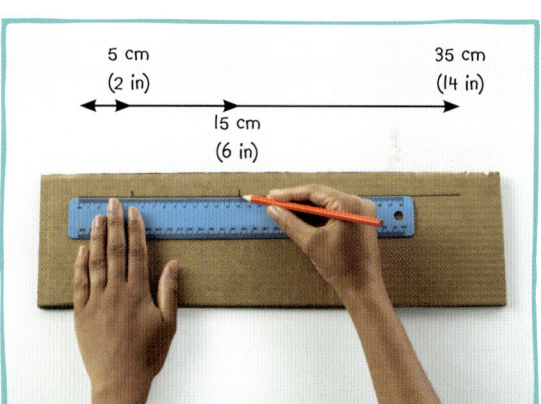

5 cm (2 in)   35 cm (14 in)

15 cm (6 in)

1 Use your ruler to draw a 35 cm (14 in) line on a piece of cardboard, near the top. Then make pencil marks 5 cm (2 in) and 15 cm (6 in) from the left end of the line.

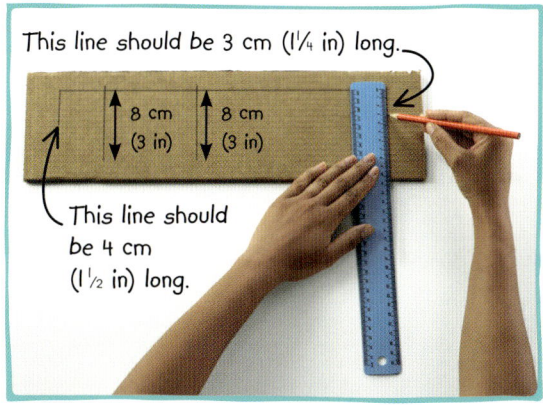

This line should be 3 cm (1¼ in) long.

8 cm (3 in)   8 cm (3 in)

This line should be 4 cm (1½ in) long.

2 Now add four vertical lines: a 4 cm (1½ in) line at the left end, two 8 cm (3 in) lines from your pencil marks, and a 3 cm (1¼ in) line on the right.

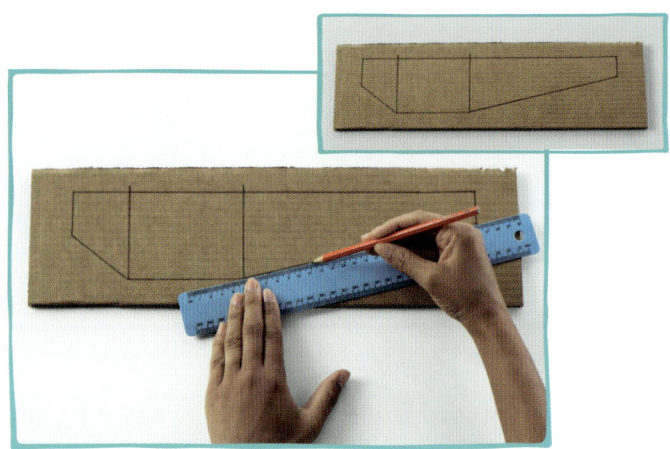

Make pencil marks on the second piece too.

3 Join up the ends of the four vertical lines. The shape you've drawn will form one side of the crane's jib.

4 With scissors, cut out the shape and then use it as a template to make an identical shape on another piece of cardboard.

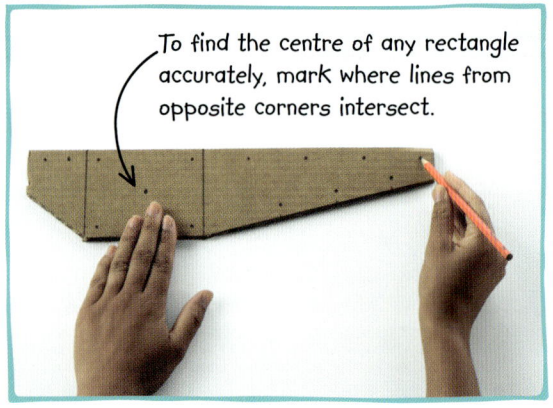

To find the centre of any rectangle accurately, mark where lines from opposite corners intersect.

Be careful of the sharp ends of the toothpicks.

Place adhesive putty under the cardboard to protect the table and your fingers.

5 On one of the shapes, add a dot in the middle of the rectangular section. Then draw dots at regular intervals along the edges as shown above.

6 Stack the two pieces of cardboard together. Using the dots as a guide, carefully push toothpicks through both pieces of cardboard.

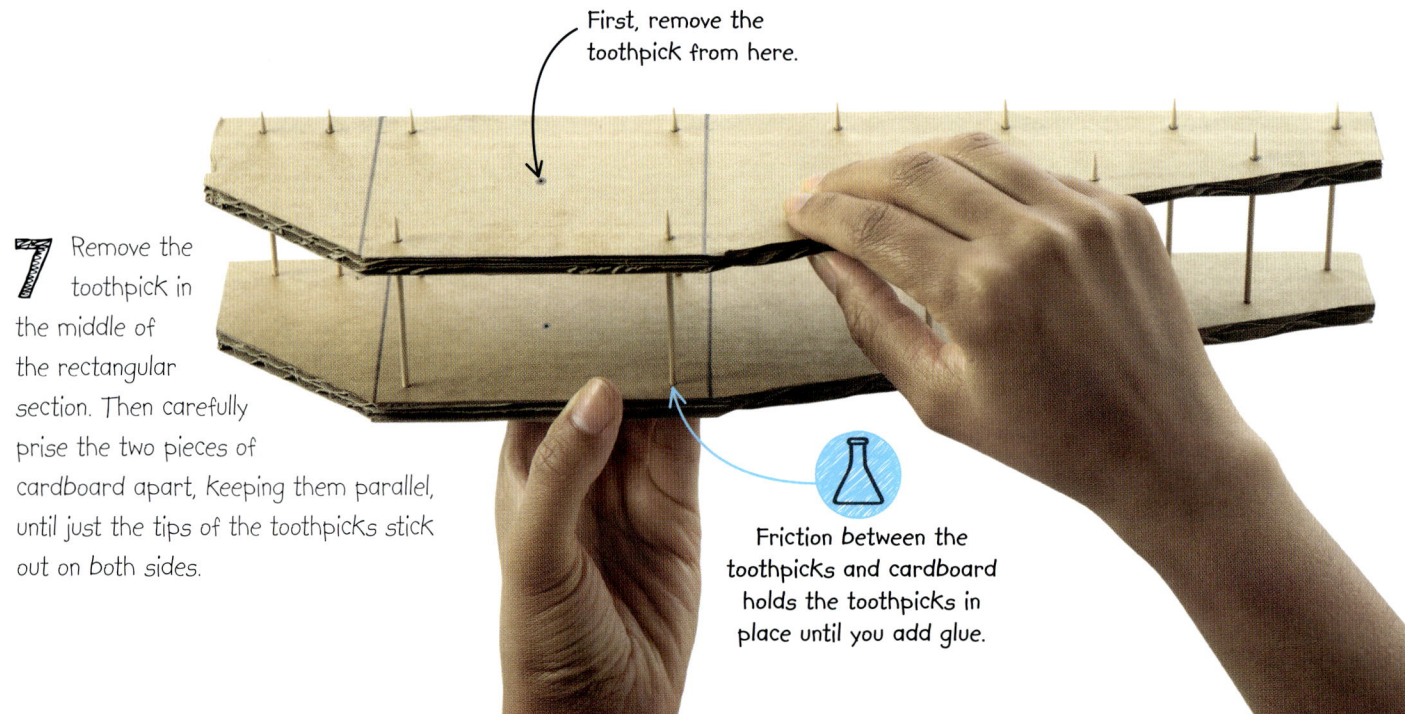

First, remove the toothpick from here.

7 Remove the toothpick in the middle of the rectangular section. Then carefully prise the two pieces of cardboard apart, keeping them parallel, until just the tips of the toothpicks stick out on both sides.

Friction between the toothpicks and cardboard holds the toothpicks in place until you add glue.

Before gluing, check you have removed the toothpick from the rectangular section.

8 Dab glue on the tips of the toothpicks. Do this first on one side and let the glue dry. Then turn the jib over and do the other side.

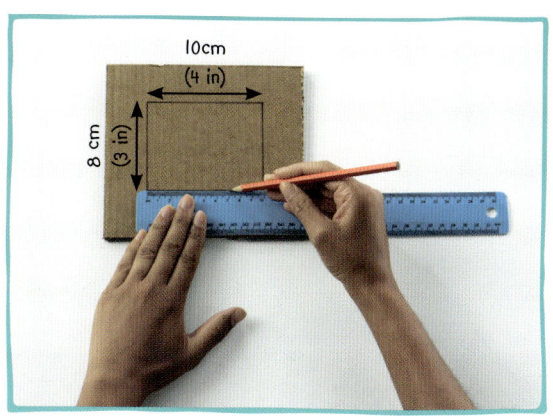

9 To make the base of your jib, draw a rectangle on a small piece of cardboard that is 8 cm by 10 cm (3 in by 4 in).

10 Turn your jib upside down and put glue on the short horizontal edges. Press the cardboard rectangle into place and let the glue dry.

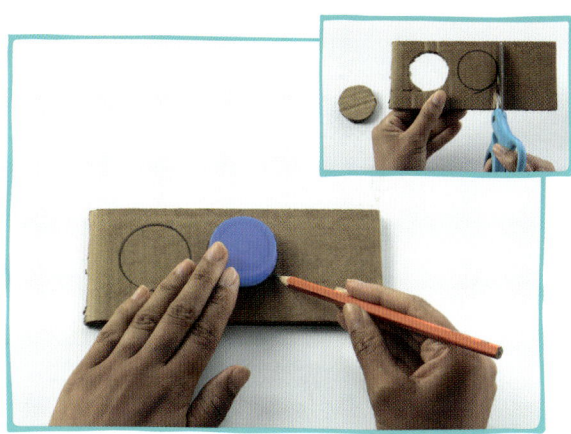

11 To make your crane's handle, draw two identical circles using the bottle cap as a guide. Then carefully cut out the circles.

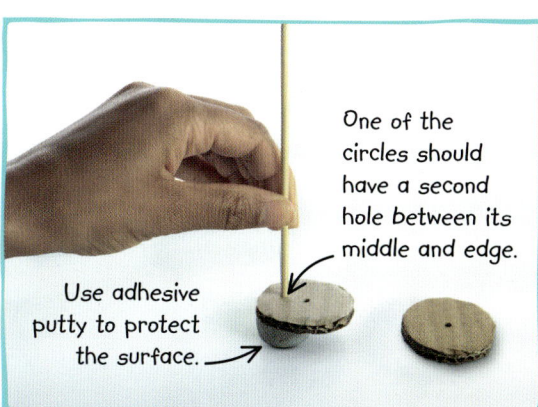

One of the circles should have a second hole between its middle and edge.

Use adhesive putty to protect the surface.

12 Use a wooden skewer to poke a hole in the middle of each circle. Then, in one of the circles, make an extra hole halfway between the middle and the edge.

Ask an adult to help you if you get stuck.

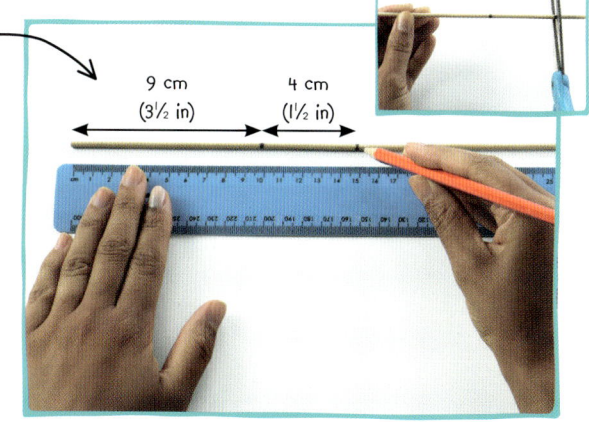

13 Measure and cut the skewer into a 9 cm (3½ in) length and a 4 cm (1½ in) length, by first scoring it with scissors, then snapping it.

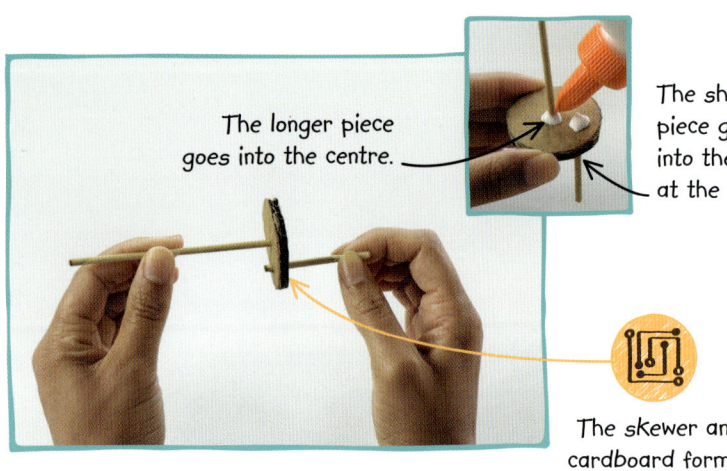

The longer piece goes into the centre.

The shorter piece goes into the hole at the side.

14 Push the two pieces of wooden skewer into the cardboard circle that has two holes. Glue them in position on both sides.

The skewer and cardboard form a device called a crank, which you will use to lift the load.

The second cardboard circle attaches to the other end of the skewer on the other side.

15 Push the long piece of skewer through the hole in the middle of the jib's rectangular section and out the other side. Glue the second cardboard circle to the other end of the skewer.

16 Cover both ends of every toothpick with a small ball of adhesive putty, to ensure that no sharp points poke out from the jib.

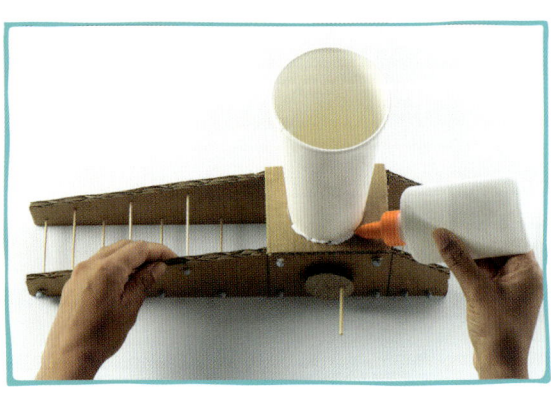

17 Turn the jib upside down and glue the bottom of one of the paper cups to the base. Wait for the glue to dry.

18 Paint your jib and paper cup. Use whatever colours you like. We've chosen yellow for the main structure and grey for the toothpicks and balls of adhesive putty.

Apply two coats of paint for a good finish.

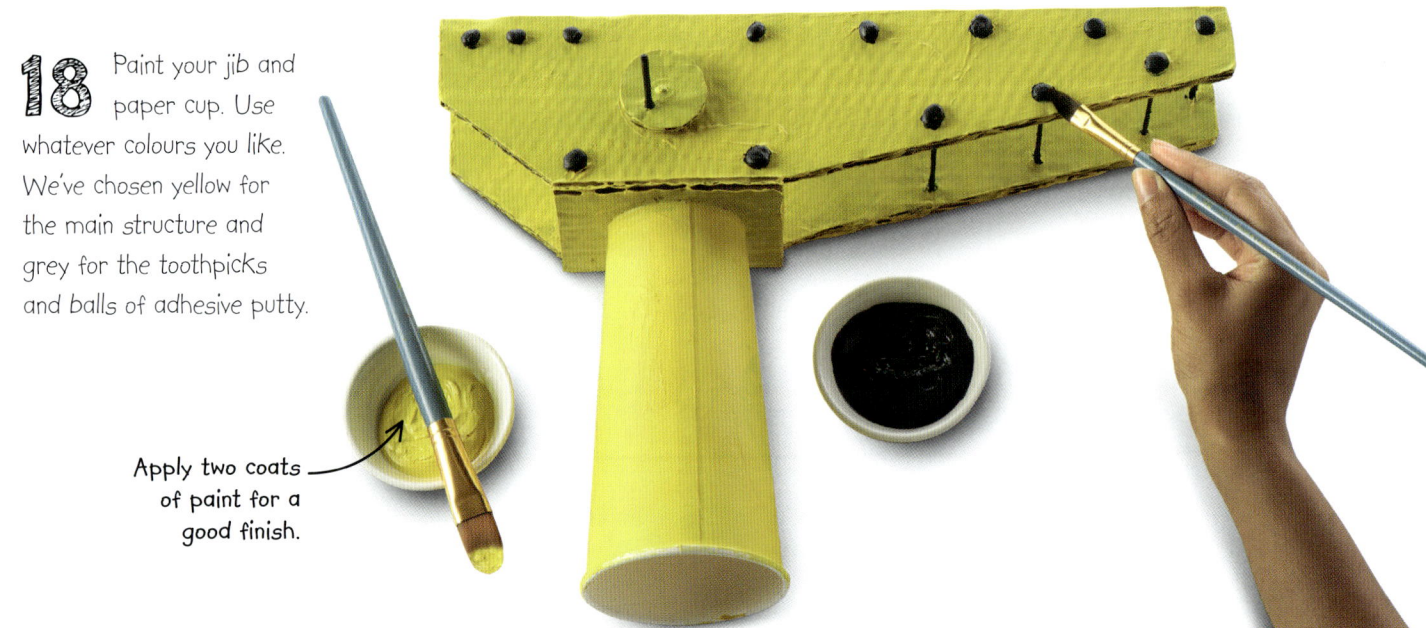

Tie a knot in the middle of the crank here.

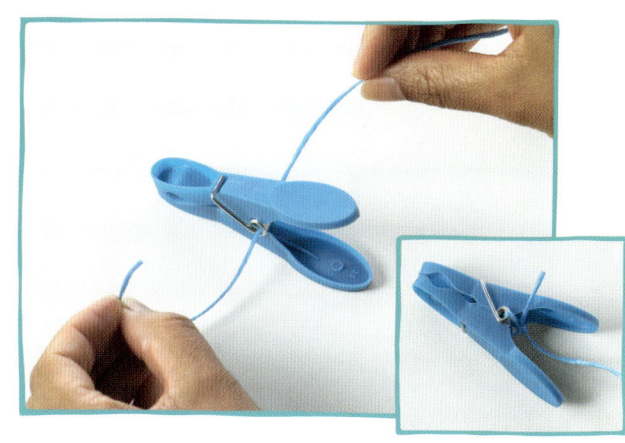

⚙ Adding some tape or a dab of glue will help prevent the string from slipping as the crank turns.

19 Cut 1 m (3 ft) of string and tie one end to the middle of the crank. Thread the other end between the two rows of toothpicks, as shown.

20 Carefully poke the long end of the string through the spring in the peg. Tie a knot to hold it in place.

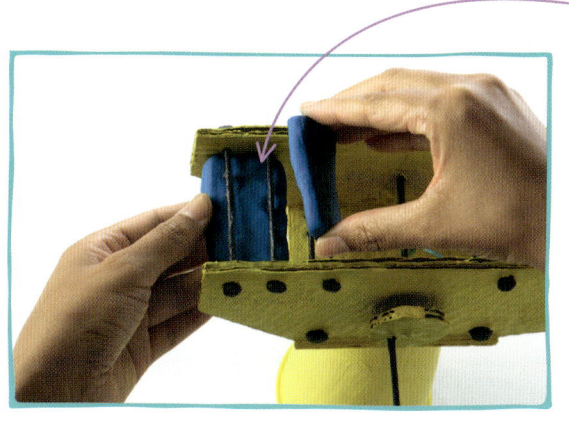

⚙ The slabs of modelling clay form a counterweight – a weight that helps balance the load.

21 Make two thick slabs of modelling clay and sandwich them together over the toothpicks at the rear of the jib.

22 Now paint the flowerpot, which will form the crane's heavy base.

⚙ The heavy, sand-filled base keeps the crane stable.

23 Take the plastic bottle filled with water and paint it with two coats of paint.

24 Stand the bottle in the flowerpot on top of the saucer and pack sand around it.

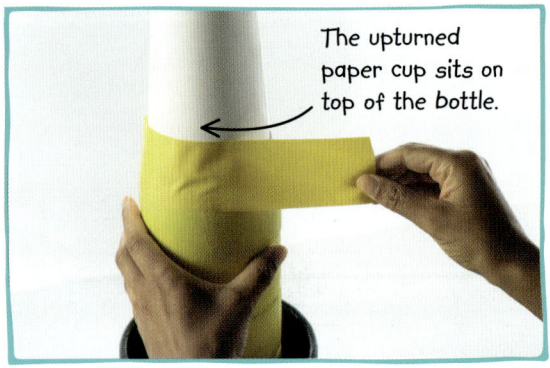

The upturned paper cup sits on top of the bottle.

25 Turn the second paper cup upside down, place it over the bottle, and secure in place with strong tape.

26 Place the marbles inside the lip of the base of the upturned cup. Make sure that there is enough room for the marbles to move a little.

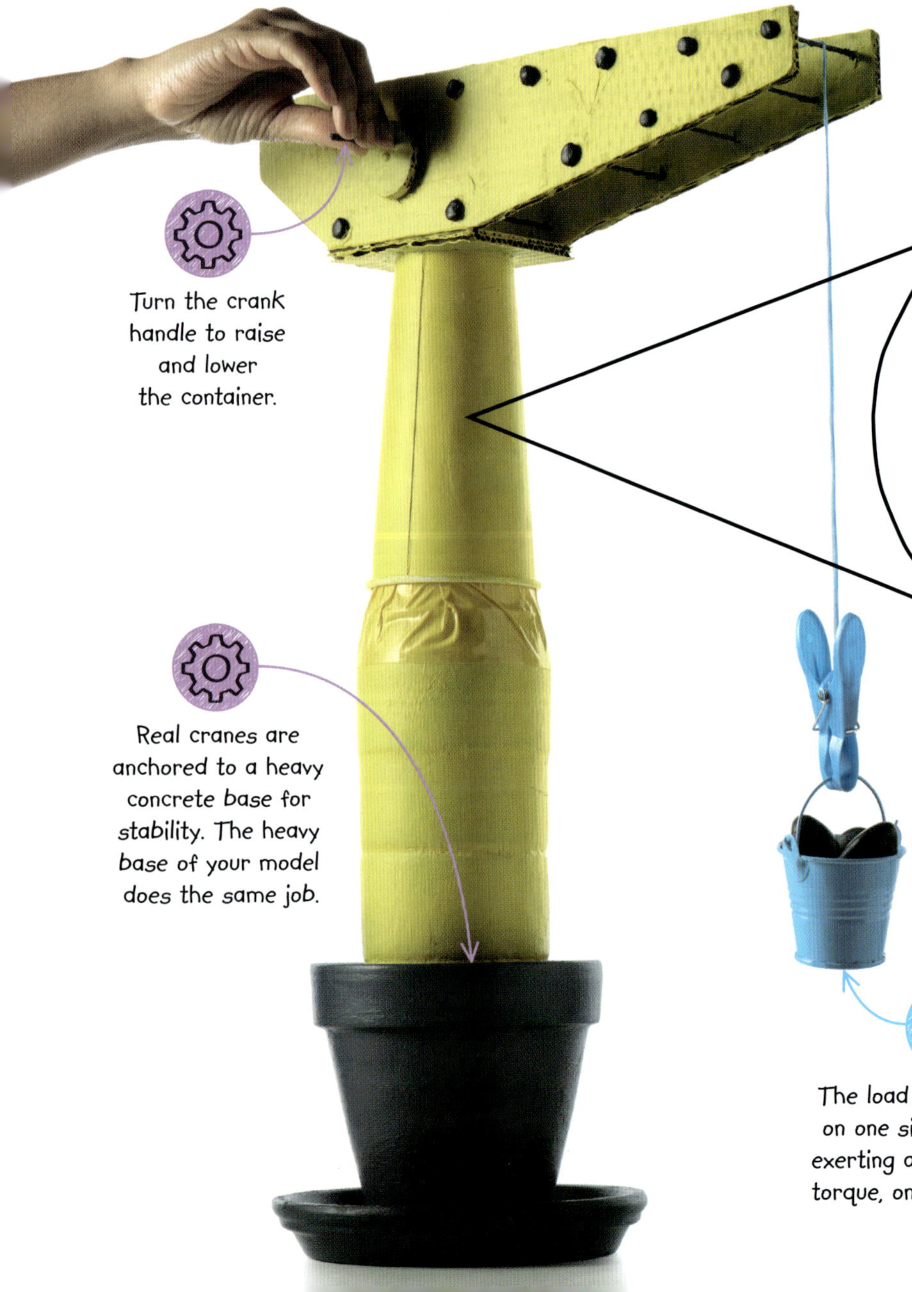

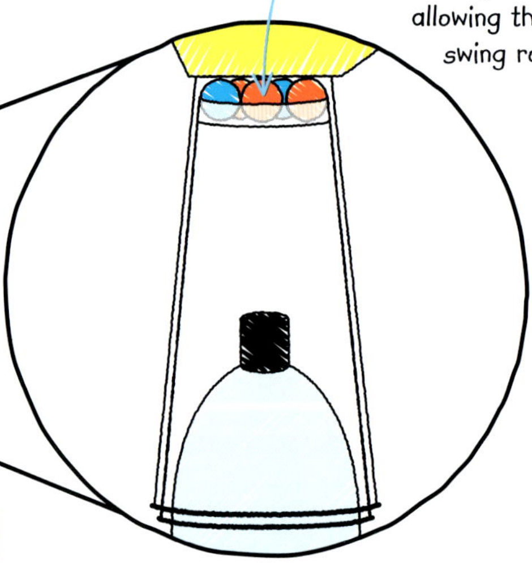

Marbles between the cups act as bearings, reducing friction and allowing the jib to swing round.

Turn the crank handle to raise and lower the container.

Real cranes are anchored to a heavy concrete base for stability. The heavy base of your model does the same job.

27 Mount the crane's jib on the tower by placing the painted cup on top of the marbles. Attach the container with a load inside it to the peg and raise or lower it by turning the handle.

The load pulls downwards on one side of the crane, exerting a turning force, or torque, on the whole crane.

TAKE IT FURTHER

See how heavy a load your crane can lift without toppling over. Try threading the string so it hangs closer to the crane's tower – can the crane lift more now? What happens if you make the wheel of the crank handle *bigger*, so that your handle moves in a bigger circle? Why not try scaling up the design, so that you can lift a heavier load? Or see what happens if you increase or decrease the weight of the counterweight? Perhaps you could use skewers instead of toothpicks, and use a double thickness of cardboard, for extra strength.

To lift heavy loads, reposition the string so it hangs closer to the tower.

HOW IT WORKS

Tower cranes can lift huge loads without toppling because they can control how far along the jib the load is positioned. Any load pulls downwards on the jib, creating a turning force, or torque. The further along the jib a load is, the greater the torque (the torque equals the load's weight multiplied by its distance from the tower). Large loads are lifted close to the tower, and small loads are lifted further out. As a result, both produce a similar torque, which is roughly balanced by the counterweight. They don't need to be perfectly balanced because the crane is also anchored to the ground.

REAL WORLD: ENGINEERING CONSTRUCTION CRANES

In a real tower crane, a steel cable hangs from a mobile trolley that can move back and forth along the jib. By varying the position of the trolley, the operator can change the torque created by the load. In your crane, threading the string over different toothpicks does the same thing. A tower crane can lift up to about 20 tonnes – as much as 20 cars.

Distance A   Distance B

Weight A (counterweight)

Weight B (load)

The crane is perfectly balanced when weight B x distance B = weight A x distance A.

AUTOMATON

An automaton is a mechanical figure that appears to move of its own accord. In reality, automata are powered by hand, clockwork, or any source of moving energy – also known as kinetic energy. Automata date back more than 2,000 years and were often built to entertain audiences. In this project, you can make a shark automaton that swishes its tail and opens its jaws. These movements are controlled by devices called cams and cranks, which are found in many kinds of machine, including car engines.

The jaw moves up and down.

This wooden skewer forms a push rod – a straight piece that is pushed up and down.

Twisting this skewer round transfers energy to the cam, which transfers it to the shark's mouth, making it open and close.

This circular piece of cardboard is a cam. It rotates off-centre on the skewer, causing the push rod to move up and down.

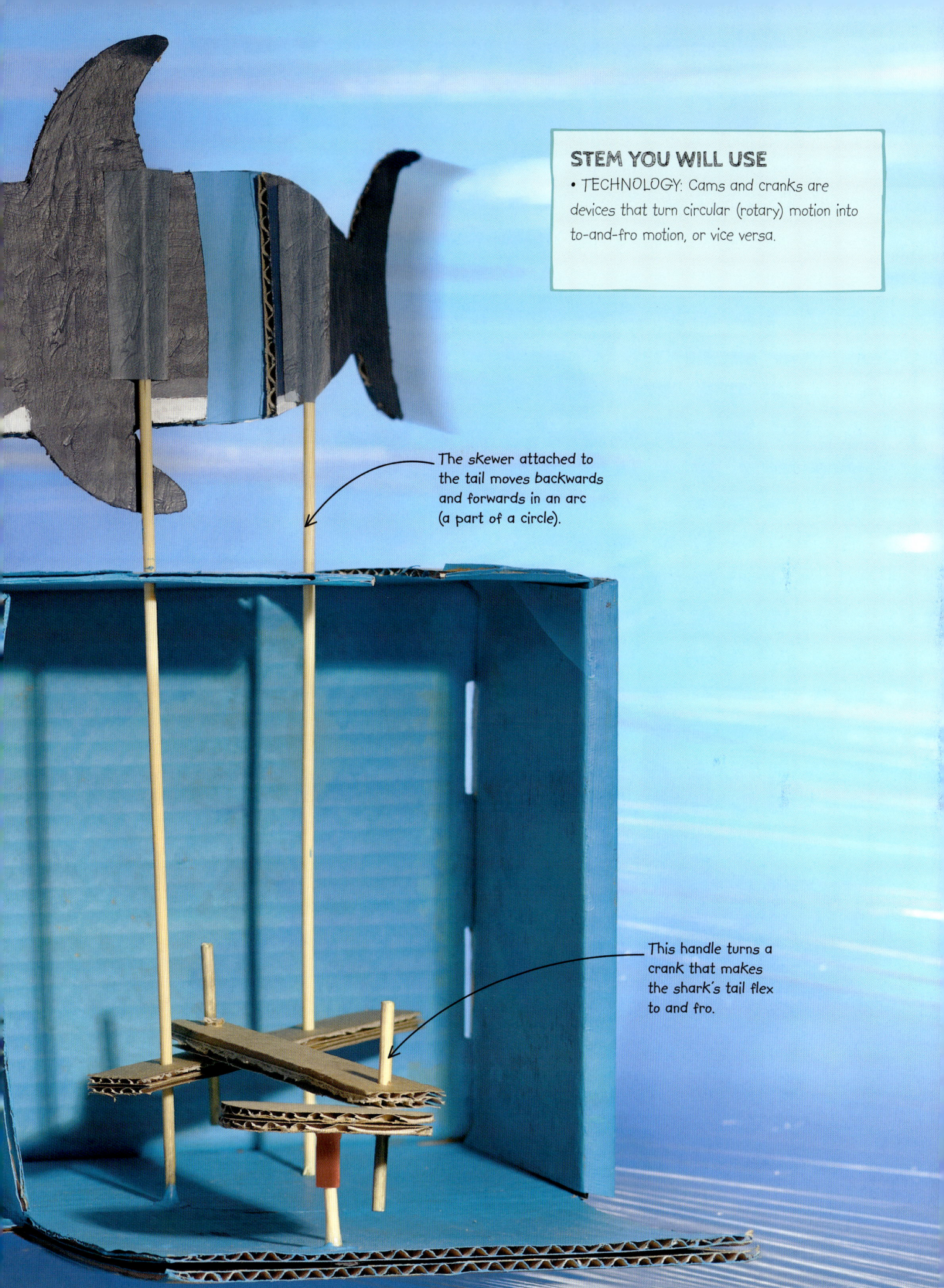

The skewer attached to the tail moves backwards and forwards in an arc (a part of a circle).

This handle turns a crank that makes the shark's tail flex to and fro.

HOW TO MAKE AN
AUTOMATON

This challenging build will take you a while. You'll need plenty of thick cardboard, as this project has lots of small pieces. You can still do this activity if you don't have a cardboard box exactly the same size as ours, but you may have to adjust some of the other pieces you cut.

Time
2 hours

Difficulty
Hard

WHAT YOU NEED

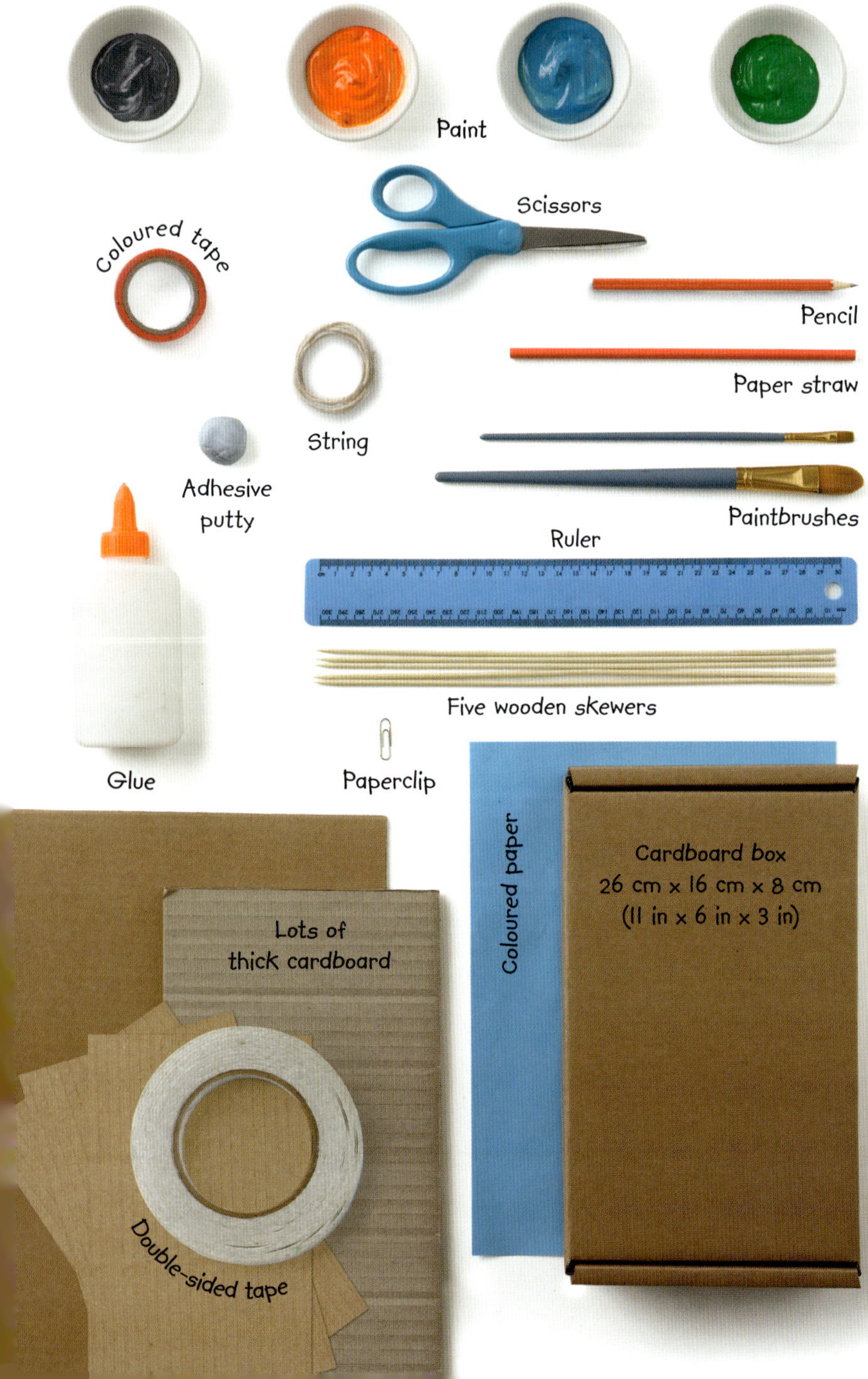

Paint

Scissors

Coloured tape

Pencil

Paper straw

String

Adhesive putty

Paintbrushes

Ruler

Glue

Five wooden skewers

Paperclip

Lots of thick cardboard

Coloured paper

Cardboard box
26 cm x 16 cm x 8 cm
(11 in x 6 in x 3 in)

Double-sided tape

1 To make the base of your automaton, draw a rectangle 20 cm by 15 cm (8 in by 6 in) on a small piece of thick cardboard.

2 On another small piece of cardboard, draw a second rectangle 12 cm by 15 cm (5 in by 6 in). Then cut out both rectangles.

3 Apply double-sided tape across the middle and around the edges of the larger rectangle. Peel the protective strip off the tape.

The base attaches to the long side of the box.

4 Stick the two rectangles together, as shown, leaving part of the large rectangle exposed. Place the long side of the box onto the exposed part.

Each mark should be 2 cm (¾ in) from the edge of the box.

The pencil line will act as a guide for where to insert the wooden skewers.

5 On the top, make three marks 2 cm (¾ in) from the edge in the middle and at both ends. Draw a straight pencil line to join them.

6 To make the mechanism for the automaton's right-hand section, draw two 12 by 2 cm (5 by ¾ in) rectangles on cardboard. Cut them out.

Hold the cardboard next to the ruler as you mark it.

7 Make three dots along the middle of one of the small rectangles, 2 cm, 4 cm, and 8 cm (¾ in, 1½ in, and 3 in) from one end.

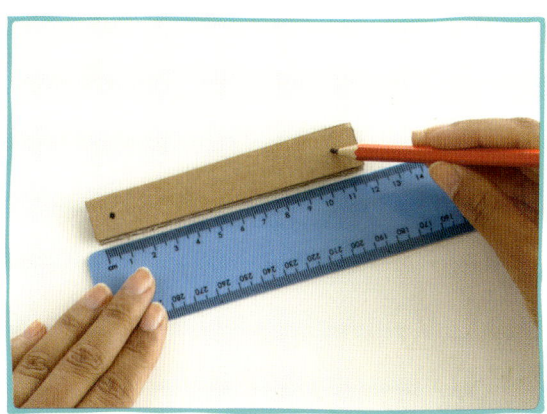

8 On the other small rectangle, make two pencil dots along the middle, 1 cm (½ in) from each end.

Use adhesive putty to protect the table.

Put adhesive putty on the other side of the card to protect your fingers.

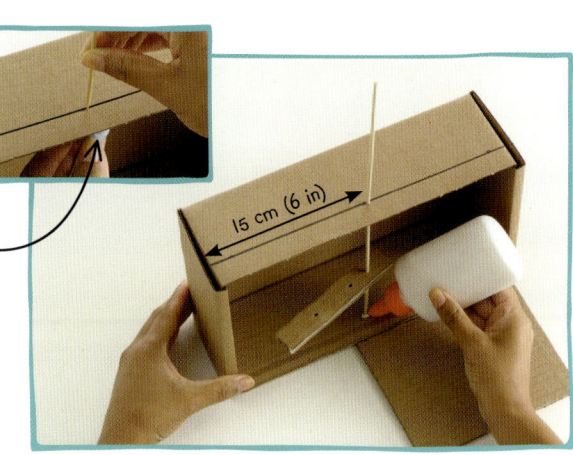

15 cm (6 in)

9 Using a wooden skewer, make holes in the dots you made on the small rectangles.

Ask an adult to help if you find this part tricky.

10 Push the skewer through the top of the box 15 cm (6 in) in from the left, then through the hole in the rectangle with three holes, as shown. Poke the skewer into the base of the box and glue it.

This is the 5 cm (2 in) piece of skewer.

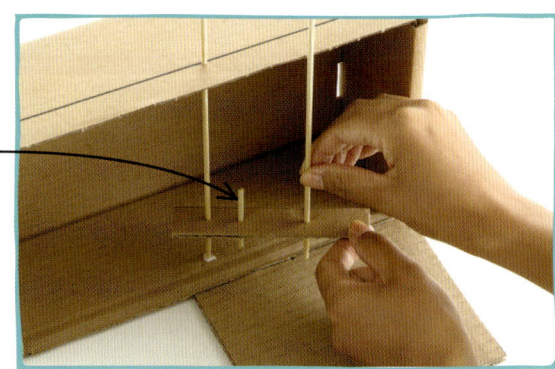

11 Take another skewer and make marks 5 cm, 10 cm, and 13 cm (2 in, 4 in, and 5 in) along it. Use scissors to score and break it at each mark.

12 Push the 5 cm (2 in) piece of skewer through the middle hole. Then place a full-length skewer through the remaining hole.

The distance between the skewer and pencil should be 6 cm (2½ in).

The handle will be placed on top of this piece of skewer.

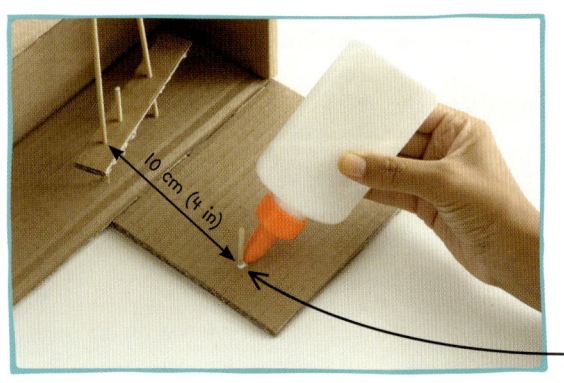

10 cm (4 in)

13 Push the 3 cm (1¼ in) piece of skewer into the base so that it is 10 cm (4 in) away from and in line with the skewer that goes through the top of the box. Add glue.

14 Use a short piece of string tied to a pencil to draw an arc around the skewer that goes through the top of the box. Cut a 1 cm (½ in) wide slit along the arc.

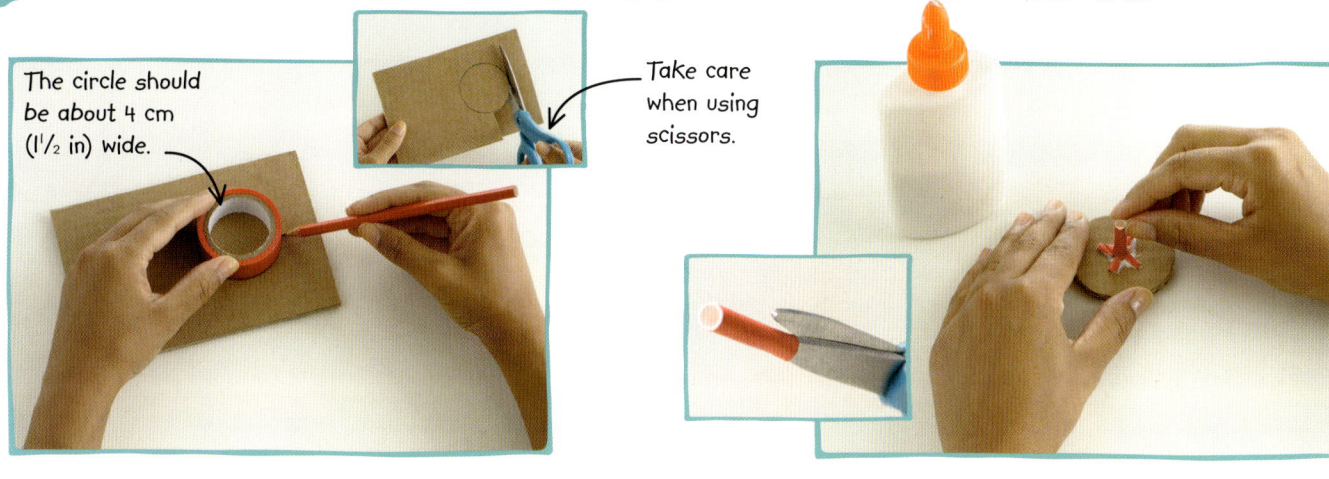

The circle should be about 4 cm (1½ in) wide.

Take care when using scissors.

15 To make the handle for the right-hand section of the automaton, draw round a roll of tape on cardboard to make a circle. Cut it out.

16 Cut a 4 cm (1½ in) length of straw. Cut four slits at one end and fold them out. Glue these flaps down onto the cardboard circle.

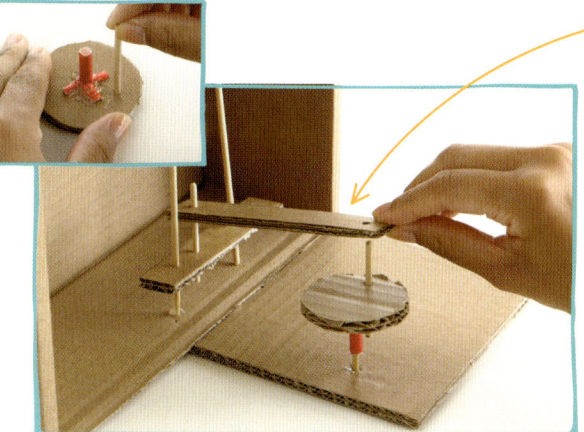

The small rectangle acts as a crank. It turns the handle's rotation movement into a back-and-forth motion.

17 Push the remaining 5 cm (2 in) piece of skewer into the handle close to the edge of the circle. Turn the handle over and place the straw over the small piece of skewer you stuck to the base. Then put the spare rectangle from step 9 over the two small pieces of skewer, as shown.

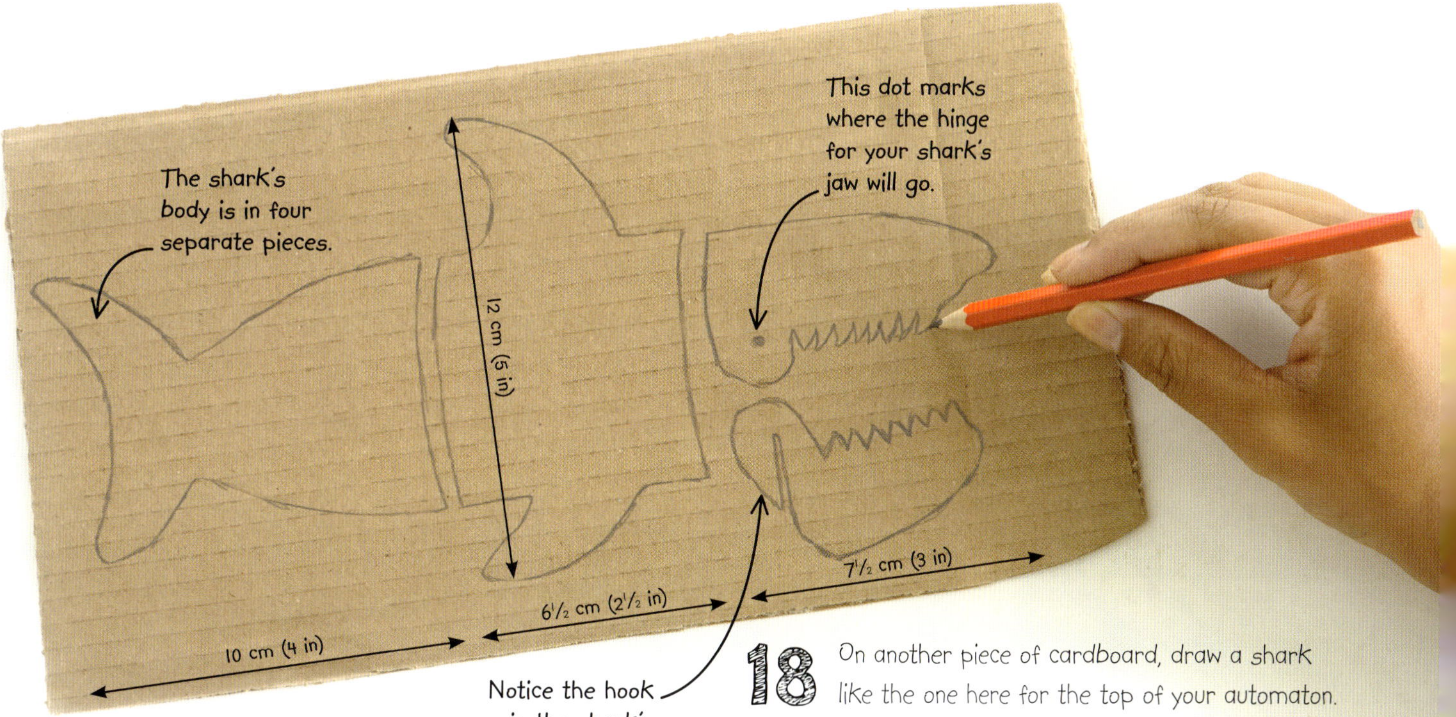

The shark's body is in four separate pieces.

This dot marks where the hinge for your shark's jaw will go.

12 cm (5 in)

10 cm (4 in)

6½ cm (2½ in)

7½ cm (3 in)

Notice the hook in the shark's lower jaw.

18 On another piece of cardboard, draw a shark like the one here for the top of your automaton. The shark should be about three-quarters the length of your box. Cut out the pieces of the shark.

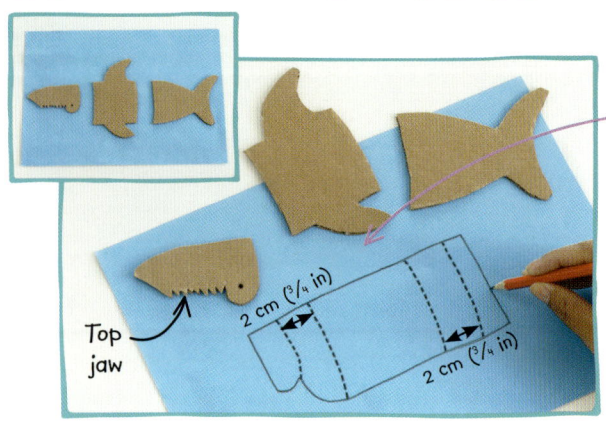

This paper backing will allow the shark's body to move freely once it is attached to the skewers.

Top jaw

2 cm (³⁄₄ in)

2 cm (³⁄₄ in)

19 Place the three shark pieces shown above on coloured paper, with gaps of about 2 cm (³⁄₄ in) between them. Draw around them and join the gaps to create the outline shown.

Don't put double-sided tape in the gaps.

20 Cut out the paper shape. Attach the pieces of the shark's body to the paper using double-sided tape. Be sure to leave 2 cm (³⁄₄ in) gaps between the top jaw, the middle, and the tail.

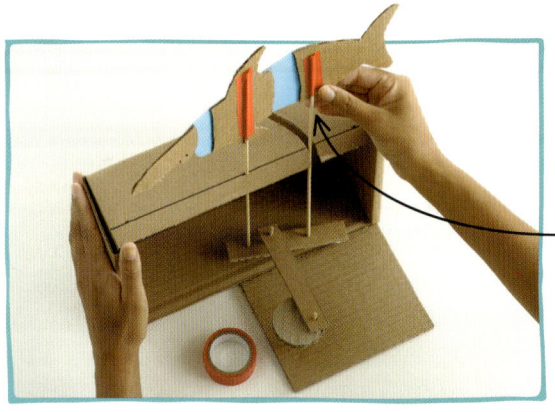

Attach the skewers to the shark's body using coloured tape.

21 Use tape to attach the shark's body and tail to the two skewers that are sticking up from the box. Snip off any excess bits of skewer that stick out above the shark.

22 You now need to add a wall to divide your automaton into two sections. To do this, measure and cut a rectangle that matches the depth and height of your box.

9 cm (3½ in)

This section will house the mechanism for your shark's jaw.

23 Glue the rectangle of cardboard 9 cm (3½ in) from the left side of the box. Make sure it doesn't prevent the mechanism on the right from moving freely back and forth.

This circle will form part of a cam – a device that turns circular motion into to-and-fro motion.

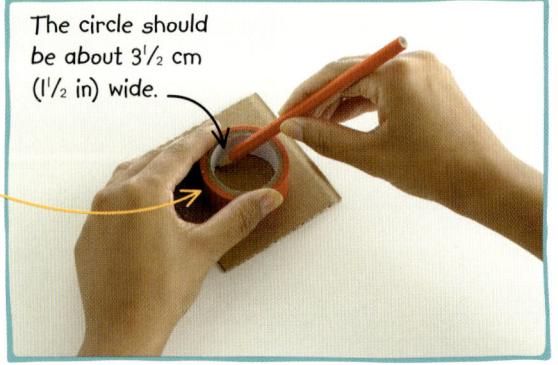

The circle should be about 3½ cm (1½ in) wide.

24 To make the mechanism for your automaton's left-hand section, draw a small circle on a piece of cardboard, using the inside of the coloured tape as a guide, and cut it out.

The diagonal lines help you work out the centre point.

In order to make the cam work, the skewer is off-centre.

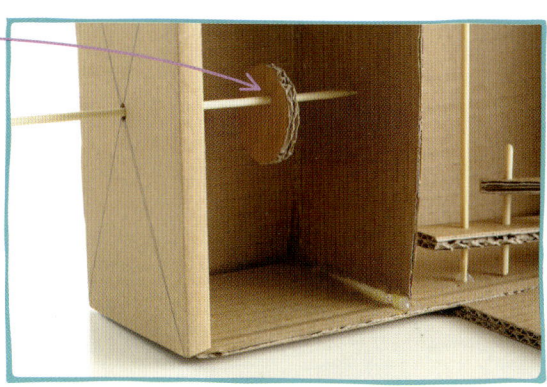

25 Turn the box on its right side and draw diagonal lines across the top end to find the centre. Push the pointed end of a skewer through the point where the diagonal lines meet.

26 Turn the box back on its base. Push the skewer through the circle about 1 cm (½ in) from its edge. Then gently push the skewer's tip through the cardboard wall.

Fold along the dotted line.

This is a push rod. The cam pushes it up and down.

27 Now draw and cut out a 4 cm by 6 cm (1½ in by 2½ in) rectangle, and fold it in half lengthways.

28 Push a skewer through the centre of the folded cardboard rectangle so that it just pokes through. Add a dab of glue and let it dry.

The straw acts as a bushing – a low-friction tube that allows a shaft to turn freely inside it.

This small circle will stop the lower jaw wobbling.

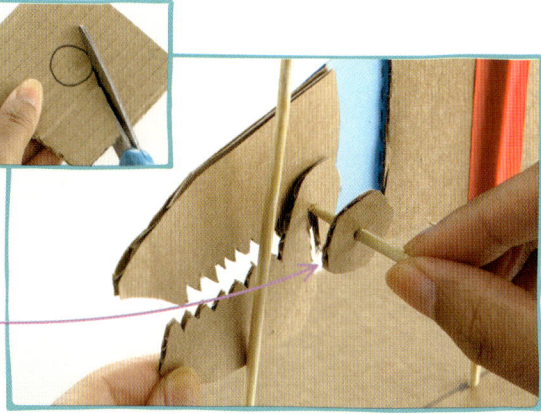

29 Make a hole on the pencil line that lines up with the middle of the jaw. Push a 5 cm (2 in) piece of straw through the hole, then insert the push rod by feeding it up through the straw.

30 To make the hinge for the jaw, cut out a cardboard circle 2 cm (¾ in) wide. Push a short piece of skewer through this and through the dot in the upper jaw. Hook the lower jaw in between.

Trim the skewer here.

Rest this cardboard on the circular cam below.

31 Tape the push rod to the shark's lower jaw, making sure the folded piece of cardboard is resting on the cam inside the box. Snip off any excess bits of skewer.

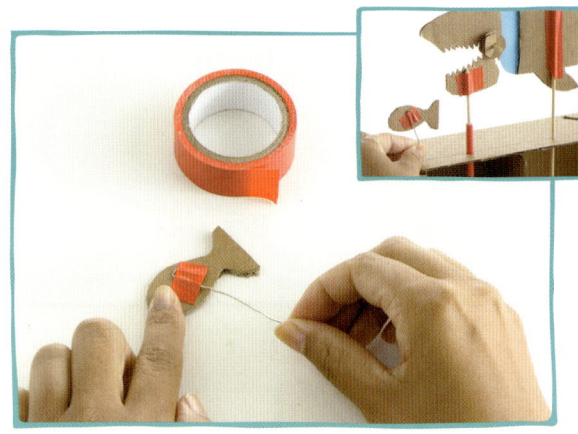

32 To decorate your automaton, draw a fish onto a small piece of cardboard, and cut it out. Straighten a paperclip, tape one end to the fish, and stick the other into the top of the box.

33 Finally, paint your automaton as you like. To make your automaton work, turn the handle to swish the shark's tail and twist the skewer at the side of the box to make the jaw move up and down.

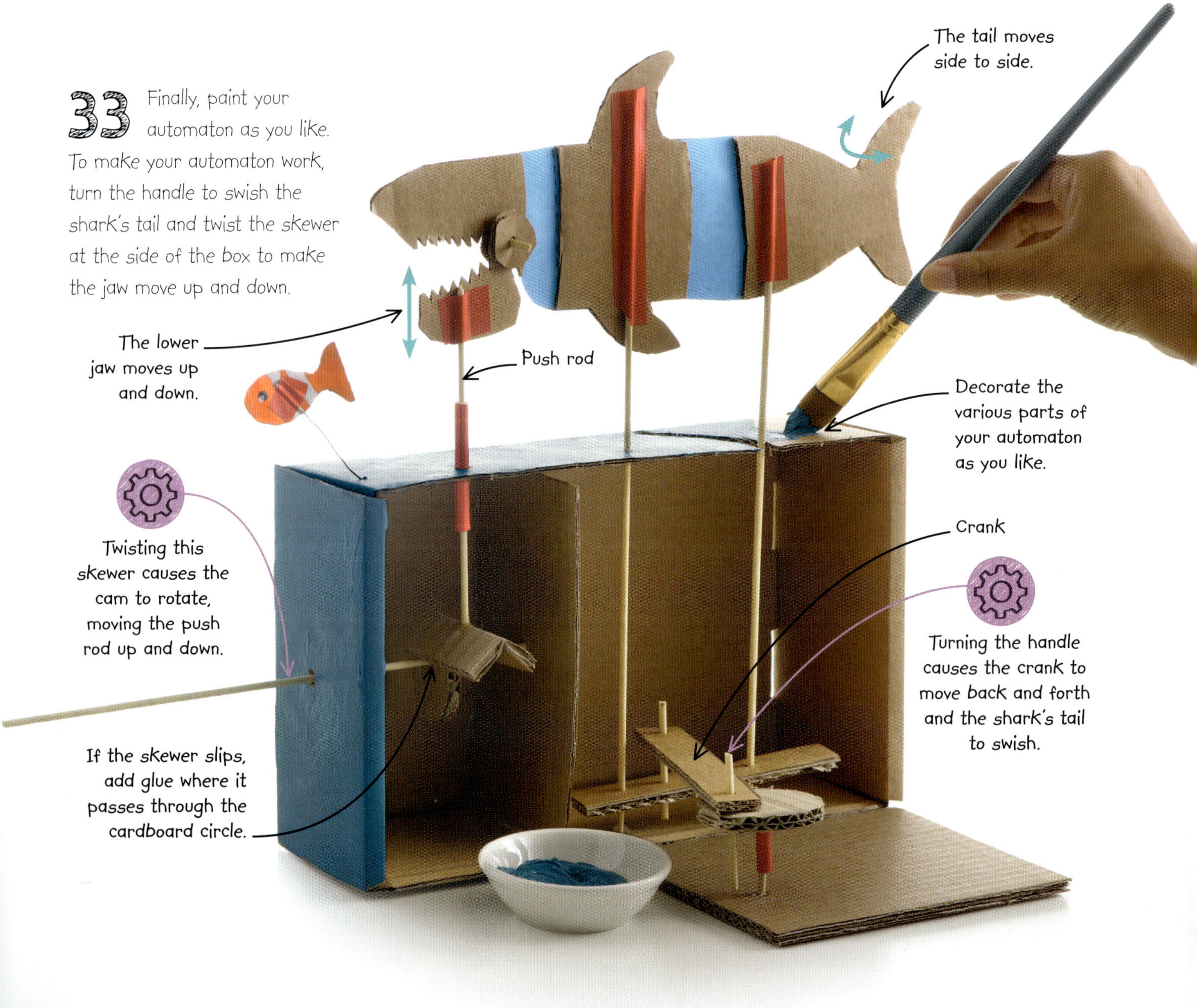

The tail moves side to side.

The lower jaw moves up and down.

Push rod

Decorate the various parts of your automaton as you like.

Twisting this skewer causes the cam to rotate, moving the push rod up and down.

Crank

Turning the handle causes the crank to move back and forth and the shark's tail to swish.

If the skewer slips, add glue where it passes through the cardboard circle.

HOW IT WORKS

When you twirl the *skewer* at the left side of the automaton, it turns the cam – the cardboard circle that is *set off-centre* on the skewer. The cam pushes a folded rectangle and the push rod up and down as it rotates, turning circular motion into up-and-down motion.

In the right-hand section the arm attached to the circle is a crank. As the handle turns, one end of the crank moves round with it, pivoting as it goes and pushing the other end in and out. A crank can turn rotation into to-and-fro motion or do the opposite, turning to-and-fro motion into rotation.

LEFT-HAND
MECHANISM

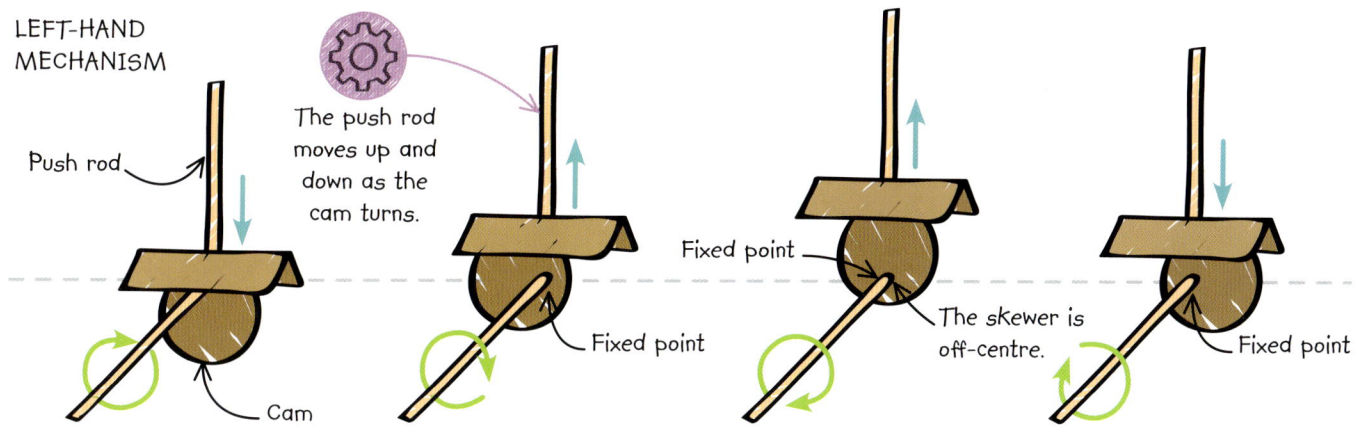

Push rod

The push rod moves up and down as the cam turns.

Fixed point

Fixed point

The skewer is off-centre.

Fixed point

Cam

RIGHT-HAND
MECHANISM

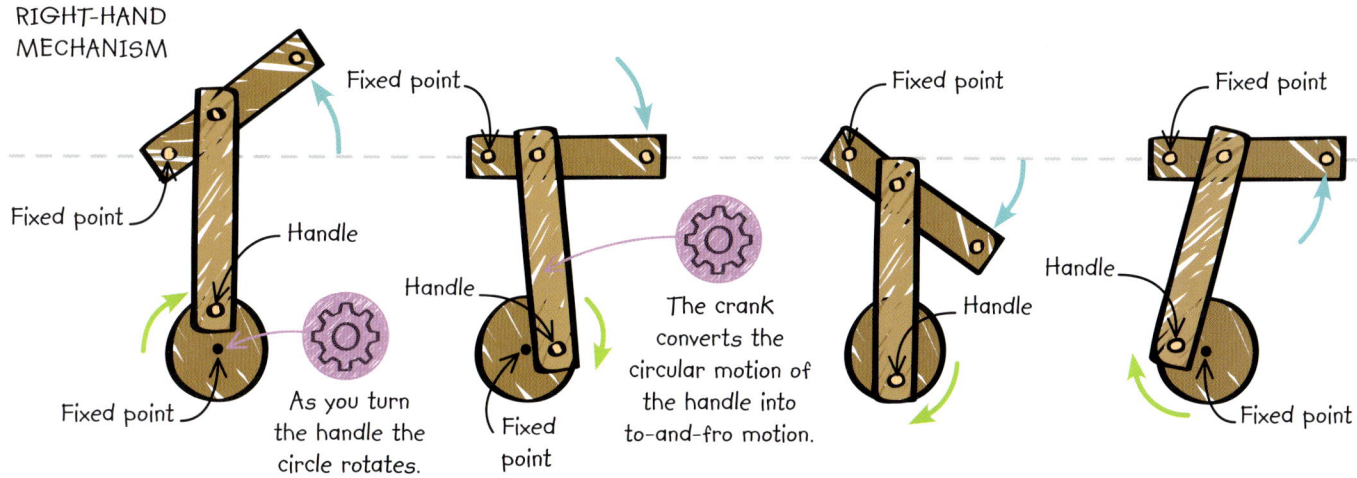

Fixed point

Fixed point

Handle

Fixed point

Handle

As you turn the handle the circle rotates.

Fixed point

Handle

The crank converts the circular motion of the handle into to-and-fro motion.

Fixed point

Fixed point

Handle

Handle

Fixed point

REAL WORLD: TECHNOLOGY
CAR ENGINE

Cranks and cams are very important components in most petrol- or diesel-powered car engines. Pistons push up or down, and cranks turn that motion into circular motion that turns the wheels. A set of cams on a shaft open and close valves that allow petrol or diesel vapour into the engine and exhaust gases out at just the right moments.

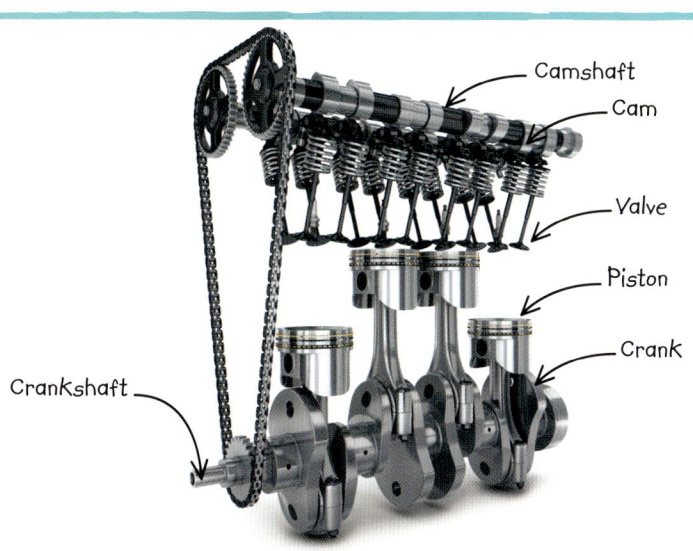

Camshaft

Cam

Valve

Piston

Crank

Crankshaft

LIQUIDS AND REACTIONS

Getting to know the properties of everyday liquids such as water and oils is a great way to learn about science. In this chapter, you'll turn your kitchen into a chemistry lab. You'll investigate how to keep liquids cool, as well as how to warm them up, and experiment with chemical reactions using a variety of household substances, including vinegar and cabbage juice. You'll also explore the science behind gels and smells by creating your own air freshener.

The temperature inside your body is around 37°C (99°F). When your surroundings are colder than that, you lose heat through your skin.

With a layer of oil between your skin and the cold water in this activity, your hand will lose heat much more slowly.

MATERIAL INSULATION
BLUBBER GLOVE

If you swam in the icy cold waters of the oceans in the Arctic or Antarctic, your warm body would quickly lose heat. One way you could slow down that heat loss would be to wear a suit that contained a layer of oil or fat – the suit would work just like the fatty *blubber* that whales, dolphins, and seals have to protect them from the cold. Materials that protect against the cold are known as insulation.

HOW TO MAKE A
BLUBBER GLOVE

For this activity, you need to create a layer of vegetable oil or sunflower oil next to your skin before plunging your hand into icy water. You'll be doing that by making a double layer glove using kitchen bags. When you've finished with the oil, put it in the bin – don't pour it down the sink as it might cause a blockage!

Time
15 minutes

Difficulty
Easy

WHAT YOU NEED

Vegetable or sunflower oil

Scissors

Duct tape

Two small kitchen bags (with seal tops)

Large bowl or bucket of icy water

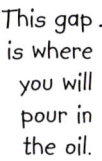

Timer

Funnel

1 Turn one of the small kitchen bags inside out, by reaching inside it and pulling the bottom out through the top. Straighten the bag out.

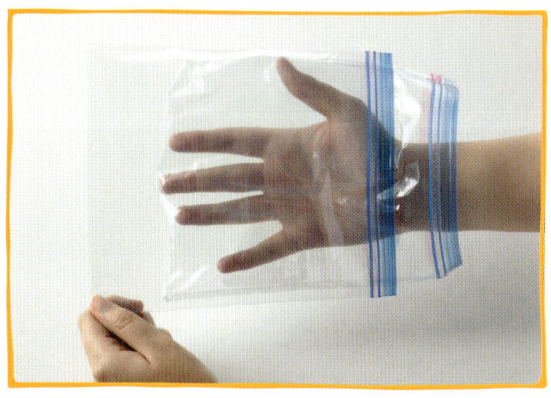

2 Put your hand inside the inside-out bag and then put it into the other kitchen bag. The zip on the inner bag faces outwards and the zip on the outer bag faces inwards.

This gap is where you will pour in the oil.

3 Press the zips on the two bags together to create a seal. Make sure you leave one part unsealed.

4 Apply tape to the parts of the bags you have pressed together and fold it over to seal them together. Do not put tape over the gap you left.

Ask for help if you find it difficult to hold the bag, the funnel, and the bottle of oil.

5 Put the funnel's spout into the gap between the bags. Carefully pour in enough oil to fill the bag two-thirds full.

The tape will help prevent any oil leaking.

6 Zip the gap up, and use another piece of tape to cover it, sealing in the oil. Fold over the tape.

7 Hold one hand under the icy water, and time how long you can leave it in there before it becomes uncomfortable.

8 Wait a few minutes for your hand to return to normal temperature, then put it inside the inner bag. Now plunge your hand, inside the blubber glove, into the icy water. Time how long you can keep your hand in the water now, and compare it with the time you recorded before.

Vegetable oil is made of chemical compounds called fats, just like the fatty tissue that whales, dolphins, and seals have.

Make sure no cold water spills into the glove.

TEST AND TWEAK

Nerve endings in your skin give you your sense of hot and cold. They don't detect actual temperatures – instead, they sense the loss or gain of heat from your skin. Try this activity to see how it works.

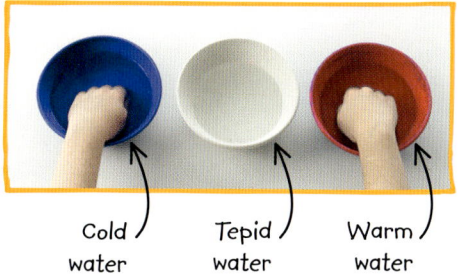

Cold water   Tepid water   Warm water

1 Fill one bowl with cold water, one with tepid water, and one with warm water. Keep one hand in cold water for a few seconds, so that it cools down, and the other in warm water, so that it warms up.

2 Then put both hands in tepid water. The hand that was in cold water will feel warm, while the hand that was in warm water will feel cold, as heat flows either to or from each hand.

HOW IT WORKS

When you plunge your hand into the icy water, heat flows quickly from your skin to the water, and you feel the cold within seconds. When you use the blubber glove, the layer of oil around your hand slows down that loss of heat, because heat flows more slowly through the oil. Materials through which heat passes slowly are called insulators. Like oil, air is a good insulator. Woolly jumpers trap lots of air, which is why they are good at keeping you warm on a winter's day.

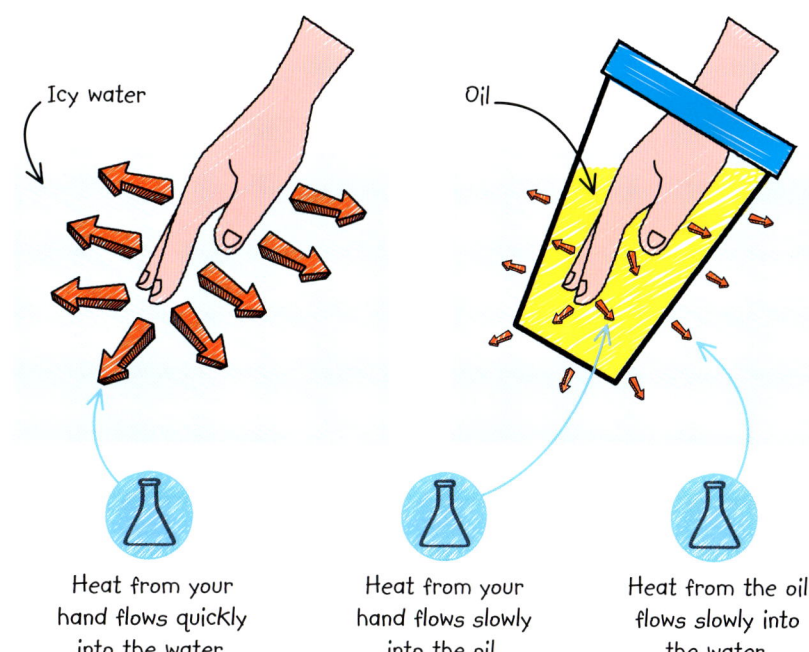

Icy water

Oil

Heat from your hand flows quickly into the water.

Heat from your hand flows slowly into the oil.

Heat from the oil flows slowly into the water.

REAL WORLD: SCIENCE
BLUBBERY ANIMALS

Whales, dolphins, and seals are mammals. That means they are warm blooded. In order to survive in the icy Arctic or Antarctic, where water temperatures are extremely cold, these animals have a thick layer of fatty tissue called blubber. Blubber slows down the loss of heat from their bodies. A blue whale's blubber layer is 30 cm (12 in) thick!

HEAT TRANSFER
THERMOS

If you're having a picnic on a hot day, it's a real treat to be able to have a cool drink. But if you take a cold drink in an ordinary bottle, the Sun's heat and the warm air will gradually warm it up. You can keep cold drinks cold for longer by carrying them in a thermos. A thermos reduces the transfer of heat, and so keeps hot drinks hot and cold drinks cold for longer.

The foil helps prevent heat from reaching or leaving the drink inside.

There's a gap between the two bottles in this thermos that reduces the transfer of heat to or from the drink inside.

HOW TO MAKE A
THERMOS

You'll make your thermos with two *bottles* – a glass one to hold the drink and a larger plastic one. It might *be* a bit difficult to get bottles that nest together well, but try to find a glass bottle that is slightly smaller than the plastic one – it is important that there is a gap *between* the two bottles.

Time
30 minutes

Difficulty
Medium

STEM YOU WILL USE

• SCIENCE: Substances cool down when they lose heat to their surroundings.
• ENGINEERING: Air gaps and reflective surfaces such as metals are often used to reduce the transfer of heat.

WHAT YOU NEED

Ruler

Two glasses

Jug and iced water

Funnel

Scissors

Coloured tape

Adhesive putty

Foil

Glass bottle

Plastic bottle

Thermometer

1 With scissors, carefully make a hole about 1 cm (½ in) down from the bottle cap. Cut the top off the bottle. Ask an adult if you find this tricky.

2 Cut around the middle of the bottle. You should end up with three parts: the very top, with the cap, and the top and bottom halves of the bottle.

3 Stick a lump of adhesive putty on top of the bottle cap, then push the upside-down cap into the bottom of the plastic bottle.

4 Cut a piece of foil about 30 cm (12 in) long and wrap it around the glass bottle, including the bottom, so that no glass is visible.

5 Put the glass bottle on top of the upturned bottle cap inside the plastic bottle. Slip the top half of the plastic bottle over. Then tape the two halves back together.

6 Tape the neck of the plastic bottle to the neck of the glass bottle to create a seal.

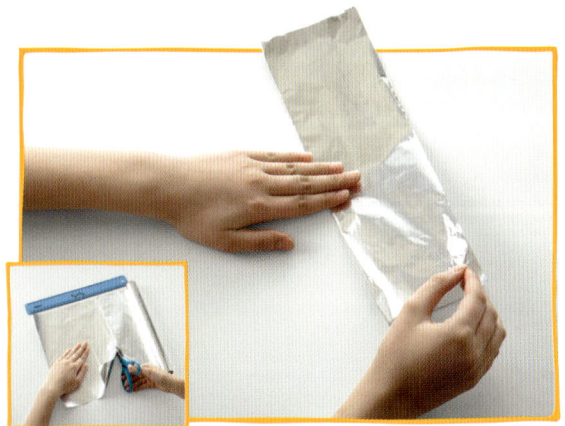

7 To insulate the top of your thermos, cut a piece of foil about 20 cm (8 in) long. Fold the foil in half lengthways, and then in half again to create a thick piece.

8 Wrap the thick piece of foil around the bottle cap of the glass bottle.

9 Pour some iced water into a glass and set aside. Remove the cap of your thermos and pour in iced water from a jug using a funnel. Put the cap back on.

The funnel will prevent water spilling down the sides of the thermos.

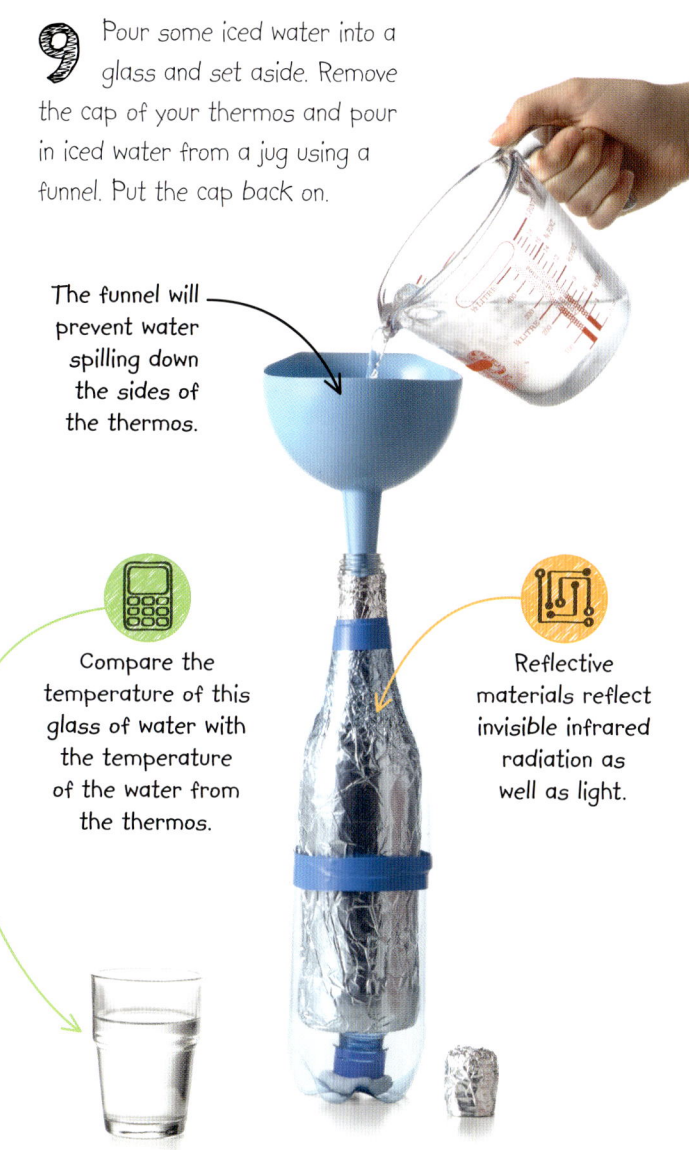

Compare the temperature of this glass of water with the temperature of the water from the thermos.

Reflective materials reflect invisible infrared radiation as well as light.

10 After about an hour, pour out some water from the thermos into another glass. Compare its temperature with the temperature of the water in the other glass.

HOW IT WORKS

Leave a glass of iced water at room temperature, and the water will slowly warm up. Heat enters the water in two ways: conduction and radiation. Conduction is the transfer of heat between two things that are in contact. Your thermos greatly reduces this effect because the air gap insulates the iced water inside the glass bottle. Invisible infrared radiation also warms the water, but these rays are reflected by the metal foil. The air and foil also prevent heat being lost from warm liquids, too.

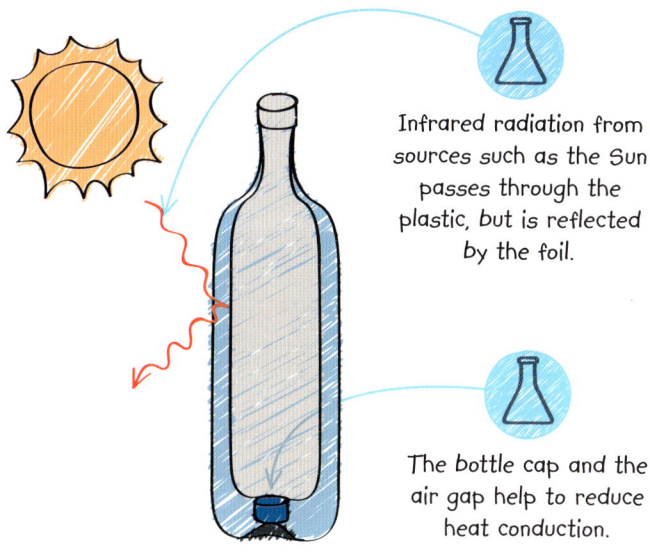

Infrared radiation from sources such as the Sun passes through the plastic, but is reflected by the foil.

The bottle cap and the air gap help to reduce heat conduction.

REAL WORLD: TECHNOLOGY

In thermos flasks you buy at the shops, the air is removed from the gap between the inner compartment and the outer surface, which creates a vacuum. The vacuum reduces heat transfer to or from the liquid inside to almost nothing. This makes it possible to keep liquids very cold or hot for long periods.

Vacuum gap

Inner compartment

The water level drops as water flows through the straw.

The pressure of the water in the cup pushes water up and over the bend in the straw.

SIPHON EFFECT
PYTHAGORAS CUP

This strange-looking contraption is named after the Greek mathematician Pythagoras, who lived 2,000 years ago. Pythagoras's cup was designed to catch out greedy people who tried to take more than their fair share of wine. When filled above a certain point, the cup drains the liquid (in this case, water) through its base. It works because a liquid will always flow from an area of high pressure to an area of low pressure – an effect known as a siphon.

STEM YOU WILL USE
• SCIENCE: Liquids and gases are both called fluids because they flow.
• TECHNOLOGY: The history of siphons goes back over 3,000 years to Ancient Egypt.
• ENGINEERING: Fluids always flow from higher-pressure areas to lower-pressure areas.

The water keeps on flowing until the cup is nearly empty.

HOW TO MAKE A
PYTHAGORAS CUP

This activity can be a bit complicated, but the surprising result is worth it, so please be patient and follow the instructions carefully. To make the Pythagoras cup, you'll need to use a lot of plastic. Make sure you recycle it once you've finished.

Time
20 minutes

Difficulty
Medium

WHAT YOU NEED

Ruler

Pencil

Bendy plastic straw

Food colouring

Scissors

Bowl of water

Plastic cup

Rubber band

Adhesive putty

Coloured tape

Dish

Plastic bottle

1 Cut around the bottle, about 7 cm (2¾ in) from the top, and keep the top part. Cover any sharp or uneven parts of the cut edge with tape.

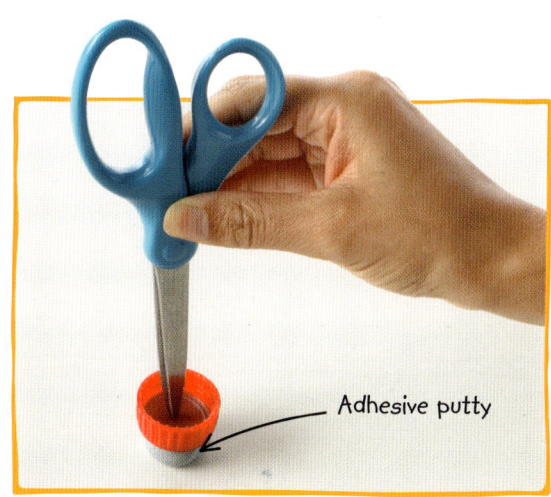

Adhesive putty

2 Remove the bottle's cap and use the scissors to make a hole in it. Place the cap on top of a lump of adhesive putty to protect the table.

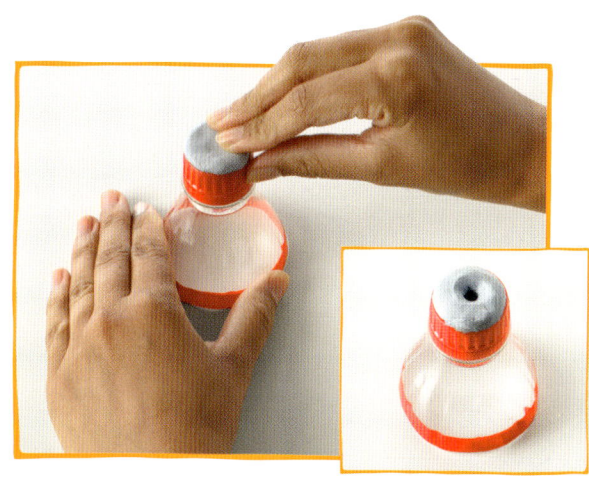

3 Press the lump of adhesive putty onto the top of the cap. With scissors, make a hole in it, in line with the hole in the cap.

The bend should be at this end.

The rubber band will ensure the fold is held close together, but not too tight.

4 Cut about 2 cm (¾ in) off the end of the straw furthest from the bendy part. Fold the straw at its bendy part and secure it in place by wrapping it with a rubber band.

5 Make a hole in the middle of the bottom of the plastic cup. Use a lump of adhesive putty underneath the cup to protect the table.

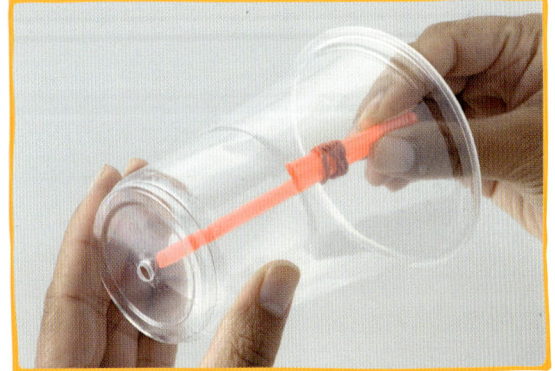

6 Feed the long end of the folded straw through the hole in the plastic cup, making sure that the folded part is inside the cup.

The adhesive putty acts as a seal to stop any leaks.

7 Push the long end of the straw through the hole in the adhesive putty and the bottle cap.

Use a pencil to pack the putty around the straw to create a seal.

8 Push the adhesive putty on the bottle cap onto the bottom of the plastic cup. Then use a pencil to pack some more adhesive putty around the straw on the inside of the top of the bottle.

9 Pour some food colouring into the water. Stand the plastic cup inside the dish. Begin pouring the water into the cup.

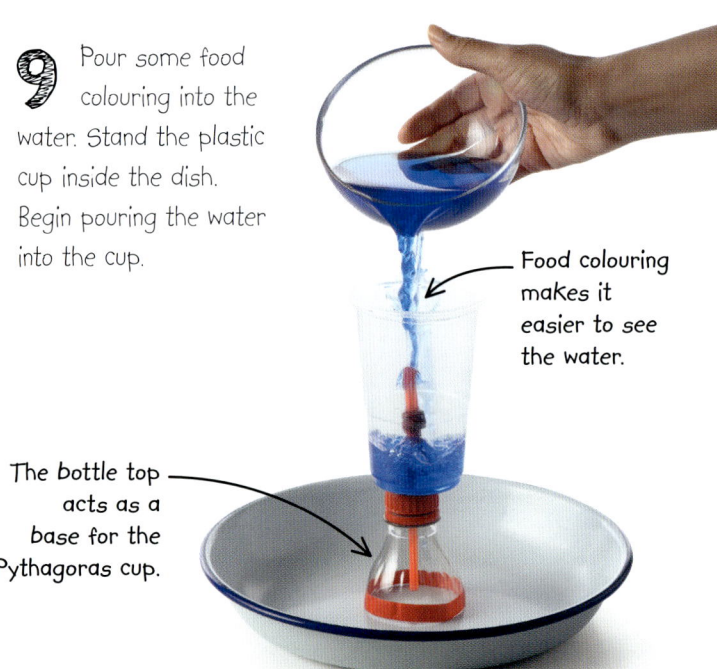

Food colouring makes it easier to see the water.

The bottle top acts as a base for the Pythagoras cup.

10 The cup will fill up until the level of the water reaches above the top of the straw. When it does, all the water will leak out!

TEST AND TWEAK

You can explore this effect with two bendy straws, two glasses (one tall, one short), and water. Make a long tube by pushing one straw into the other. Fill the tall glass with water and bend the tube so that the short end of the tube is in the tall glass. Suck a little water out through the long end to start the flow, then let the long end rest in the short glass. Water will drain from the glass until the water level goes down to the opening of the straw.

Use a piece of tape to connect the straws.

You might need a base to make your tall glass taller.

HOW IT WORKS

When you're filling the cup, water climbs up the straw, pushed by the pressure of the water, which increases as more water is poured in. When the water level in the cup is greater than the height of the straw, the pressure pushes water over the top of the bent part of the straw. The water keeps flowing, because the pressure at the bend in the straw remains lower than the pressure at the open part of the straw inside the cup. This effect is known as a siphon.

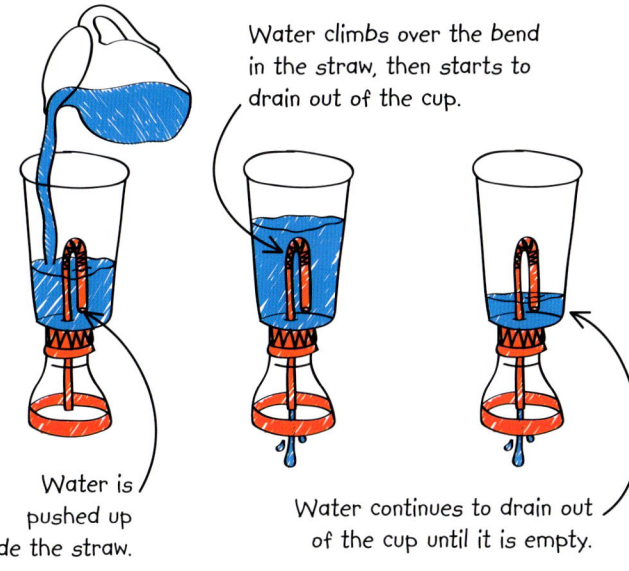

Water climbs over the bend in the straw, then starts to drain out of the cup.

Water is pushed up inside the straw.

Water continues to drain out of the cup until it is empty.

REAL WORLD: SCIENCE
TOILET FLUSH

You probably use a siphon most days – when you flush the toilet. The box at the top of the toilet is called a cistern. Inside, water fills up to just above the level of a pipe that leads down into the toilet bowl. When you turn the handle, you release water into that pipe, and all the water from the cistern rushes into the bowl.

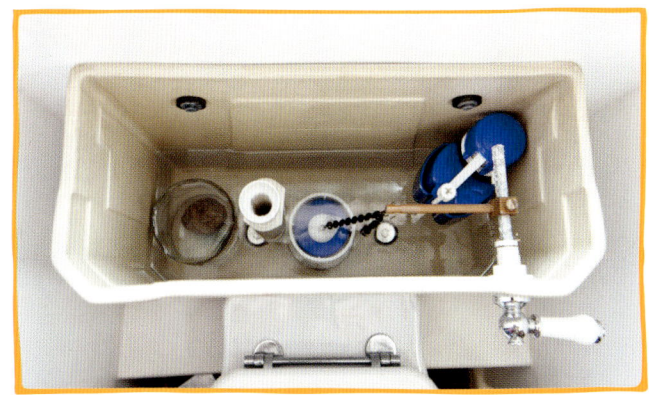

AIR FRESHENER

Natural soaps and perfumes often get their scents from essential oils – concentrated liquids (essences) extracted from plants. Some people put drops of essential oil on small dishes so the scent fills a room as the oil molecules evaporate (turn to gas). But once all the oil has evaporated, the wonderful aroma disappears. In these air fresheners, essential oils are held in a jelly-like substance called a gel. They evaporate more slowly, so they can make your room smell great for weeks.

Inside the jar is a gel – a substance that is mostly liquid but holds its shape like a solid.

For a decorative flourish, immerse cut flowers or other ornaments in your jars of smelly jelly.

STEM YOU WILL USE
• SCIENCE: Smell is caused by fragrant molecules that are volatile, which means they evaporate into the air.
• TECHNOLOGY: Preservation and gelling are two common food technologies.

A small amount of food colouring makes the gel look more attractive.

HOW TO MAKE AN
AIR FRESHENER

For this activity you need a gelling agent, which is what cooks use to make jelly. Gelling agents that work well include carrageenan and gelatine. Whatever you use, check the instructions on the packet so you make it correctly. You'll also need an essential oil. Any kind will work, but choose one you like the smell of!

Time	Difficulty	Warning
15 minutes	Easy	Hot water! Adult supervision required.

WHAT YOU NEED

Teaspoon of gelling agent
Teaspoon of salt
Jug of hot water

Glass jar
Food colouring
Essential oil

1 Add one level teaspoon of gelling agent to the glass jar.

2 Add hot water until the jar is two-thirds full. Stir with the teaspoon until the gelling agent dissolves. Take care to avoid splashing as you stir.

3 Add a few drops of food colouring, to give your mixture a bit of colour. Mix it in with the teaspoon.

Salt acts as a preservative – it prevents bacteria and fungi from growing on the gel.

 4 Now for the smelly bit. Add a few drops of an essential oil and stir the mixture again.

5 Add a teaspoon of salt and stir well until it completely dissolves.

The mixture will set in the fridge and become a gel.

6 Place the mixture in the fridge overnight to cool down and set. It's now ready to use! After 2–3 weeks, dispose of the air freshener in your food bin.

TEST AND TWEAK

The rate at which substances evaporate depends on the temperature. To investigate, place your air freshener in warm and cold places – where is the smell strongest? Add flowers, marbles or polished pebbles before the gel sets to make the air freshener look nicer. If you use flowers, dispose of your air freshener after about a week before they get mouldy.

HOW IT WORKS

Gelling agents contain long, chain-like molecules called polymers. When they dissolve in hot water and then cool down, they form crosslinks with each other, creating a three-dimensional scaffold that holds water in place and keeps its shape – a gel. The essential oil also gets trapped in the gel. As a result, it evaporates into the air more slowly, keeping your room smelling fresh for weeks.

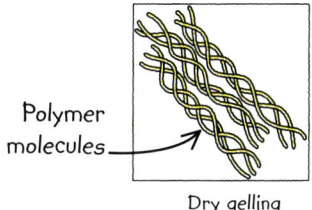

Polymer molecules

Dry gelling agent

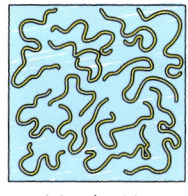

Dissolved in hot water

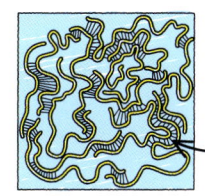

After cooling

Crosslinks between molecules

REAL WORLD: SCIENCE
CONTACT LENSES

Soft contact lenses are made from a gel containing water and plastic. As well as making them comfortable, the gel allows oxygen to pass through to the eye's surface. That's important because the eye has no blood vessels and gets all its oxygen from the air.

ACID AND BASE REACTION
BUBBLE TOWER

Turn your kitchen into a chemistry lab! At the heart of this activity is a chemical reaction between two substances: one called an acid (you'll use vinegar) and one called a base (in this case, bicarbonate of soda). The reaction produces amazing bubbles that rise through a tower of oil and then fall back down. So that the effect is even more impressive, you'll also make an indicator – a colour-changing solution that turns red in acids and blue-green in bases – using some red cabbage.

STEM YOU WILL USE
• SCIENCE: A base that dissolves in water is called an alkali.
• ENGINEERING: One liquid will always sink below and displace another less dense liquid.
• MATHS: The pH scale runs from 0 (strong acid) to 14 (strong base).

The bubbles produced by the reaction float up through oil.

The pink solution at the bottom of the vase contains an acid and a pH indicator.

A chemical reaction at the bottom of the vase produces bubbles.

HOW TO MAKE A
BUBBLE TOWER

The first thing you'll need to do is to make the pH indicator. This involves soaking pieces of red cabbage in warm water. Once you've done this, you'll add vinegar, which is an acid. Then you'll carry out the reaction in a vase, which makes it easy to watch the colourful bubbles rise and fall. When you've finished, put the oil in the bin, not down the sink.

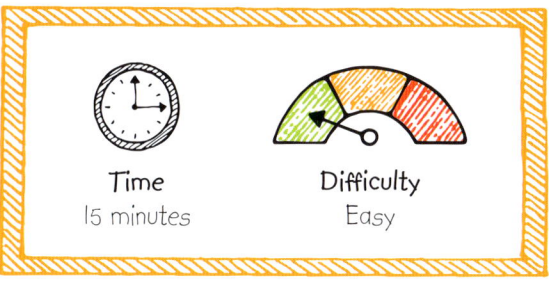

Time
15 minutes

Difficulty
Easy

WHAT YOU NEED

Measuring jug

½ red cabbage

Scissors

Small bowl

Spoon

Bicarbonate of soda

Distilled vinegar

Vegetable or sunflower oil

Vase or tall glass

Large bowl

Sieve

1 Half-fill the small bowl with warm water. Cut thin strips off the red cabbage, letting them fall into the water. Leave to stand for 10 minutes until the water turns a deep purple colour.

2 Pour the mixture through the sieve into the large bowl. Put the cabbage leaves into the food recycling bin or compost.

3 Pour 50 ml (2 fl oz) of the purple solution into the measuring jug. This solution is your pH indicator.

Your pH indicator changes colour when mixed with an acidic substance.

4 Now add 50 ml (2 fl oz) of vinegar to your pH indicator. Don't take your eyes off the jug – the solution will turn bright pink almost instantly! It turns pink *because of the vinegar's acidity.*

5 Spoon *bicarbonate of soda* into the vase or tall glass. Add enough to cover its base.

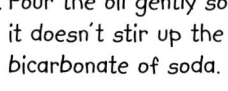
Pour the oil gently so it doesn't stir up the bicarbonate of soda.

6 Pour oil over the *bicarbonate of soda.* You need to add enough oil to fill the vase about two-thirds full.

7 Slowly pour the pink vinegar into the vase. As soon as the vinegar meets the *bicarbonate of soda,* you'll *see* red *bubbles* rise up through the oil. Watch what happens over the next few minutes. As the reaction continues, the *bubbles* will change colour as the acidity of the liquid inside them changes.

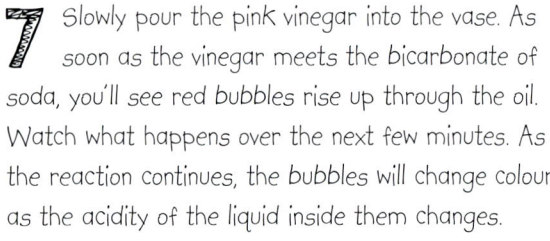

A chemical reaction between the acid and base creates bubbles of carbon dioxide gas.

The oil is less dense than the vinegar, so it floats on top.

TEST AND TWEAK

Why not try mixing your red cabbage indicator with other substances to see if it will change to any other colours. Ask an adult to help you, because some of the chemicals you find in the home can be harmful if they splash on you – especially in your eyes. Never put any household product in your mouth. Try the following: bottled water, lemonade, liquid soap, baking powder, lemon juice.

Scientists use indicators to measure how acidic or basic a solution is on a scale from 0 to 14. This is called the pH scale. Acids have a pH of less than 7, and bases have a pH of more than 7. Water, which is neither acidic nor basic, is called "neutral" and has a pH of 7. Cabbage-water indicator is red in acids, purple in neutral solutions, and turns blue-green or even green in strong bases.

HOUSEHOLD PRODUCTS

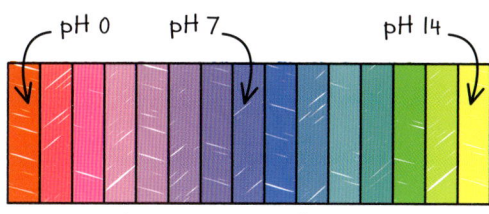

pH 0   pH 7   pH 14

RED CABBAGE pH SCALE

HOW IT WORKS

The vinegar solution is acidic. It is also more dense than the oil, so it sinks when you pour it into the vase. The acid reacts with the bicarbonate of soda, producing bubbles of carbon dioxide gas (mixed with vinegar solution), which are less dense than the oil and rise up. At the surface, the bubbles burst and any vinegar solution inside them drops down. The purple pigment in red cabbage is an indicator, a chemical that changes colour depending on how acidic a solution is. At first it is pink, but it turns blue-green as the acid is used up in the reaction.

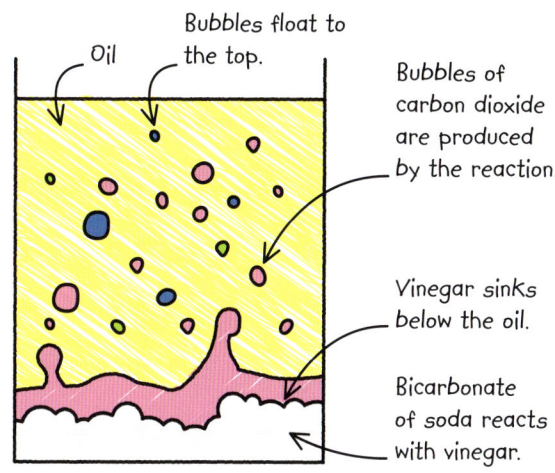

Oil

Bubbles float to the top.

Bubbles of carbon dioxide are produced by the reaction.

Vinegar sinks below the oil.

Bicarbonate of soda reacts with vinegar.

REAL WORLD: TECHNOLOGY
BAKING CAKES

You might have used baking powder to make a cake. Baking powder is bicarbonate of soda mixed with a powdered acid. When it's added to cake mixture it gets wet, and the chemicals dissolve and begin to react. Heat speeds up the reaction, producing large bubbles of carbon dioxide that make the cake rise as it bakes.

COPPER COATING

COPPER REACTIONS

Set up your own chemistry lab in the kitchen! Using just vinegar, salt, some copper-coated coins and a steel nail, you can observe some amazing chemical reactions. The coins will become shiny and new, and the steel nail will become copper-plated and change colour.

Salt and vinegar make the coins shiny.

Copper reacts with salt and acid to form a blue-green substance called copper chloride.

The steel nail is coated with copper.

A copper-coated coin reacts with salt and vinegar to produce a green solution.

HOW TO MAKE
COPPER REACTIONS

The vinegar used in these three experiments is a weak acid. It won't harm you if it gets on your skin, but if you splash some in your eyes, it will sting. If that happens, just rinse your eyes with cold water. If you spill the vinegar, or the salt, just wipe them up with a paper towel.

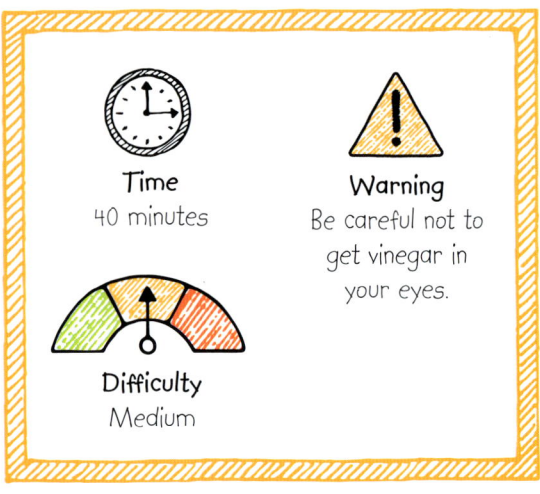

Time
40 minutes

Warning
Be careful not to get vinegar in your eyes.

Difficulty
Medium

WHAT YOU NEED

Glass bowl

Ungalvanized steel nail

10 copper-coated coins

Teaspoon

Table salt

Measuring jug

Ceramic bowl

Plastic bowl

Paper towel

Distilled vinegar

EXPERIMENT 1 – SHINY COINS

1 Put 10 coins in the glass bowl.

The coins are not copper all the way through, but they are coated with copper.

2 Pour enough vinegar into the bowl to cover the coins. Vinegar contains a chemical called acetic acid, which will react with the coins.

The chlorine in salt helps the acid dissolve the copper on the coins.

3 Now stir in half a teaspoon of salt, and leave for 10 minutes. Salt is made of the elements sodium and chlorine.

4 The coins will have turned shiny and bright. Take them out of the vinegar and salt solution. Keep them close by, in a bowl.

HOW IT WORKS

Everything around you is made of tiny particles called atoms. Atoms of different kinds (elements) join together in different ways, forming compounds. In a chemical reaction, atoms can separate, swap partners, and form new compounds. Copper coins are shiny when new but turn dull brown over time because the copper atoms react with oxygen atoms from the air to form a compound called copper oxide. Vinegar contains a compound called acetic acid, which splits in water to release positively charged hydrogen atoms (hydrogen ions). These react with the copper oxide, stripping it from the coins and revealing the shiny layer of pure copper underneath. Salt speeds up this reaction.

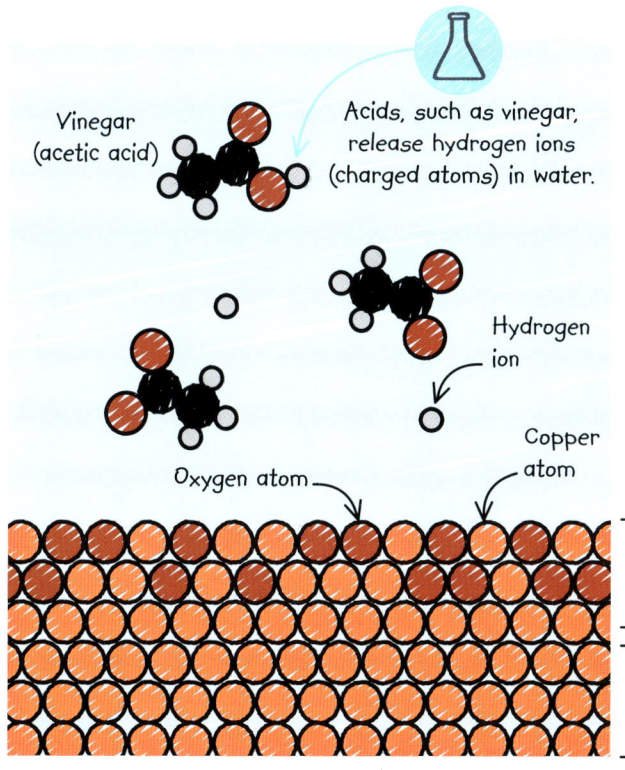

Vinegar (acetic acid)

Acids, such as vinegar, release hydrogen ions (charged atoms) in water.

Hydrogen ion

Copper atom

Oxygen atom

Copper oxide

Pure copper

EARLY STAGE OF THE REACTION

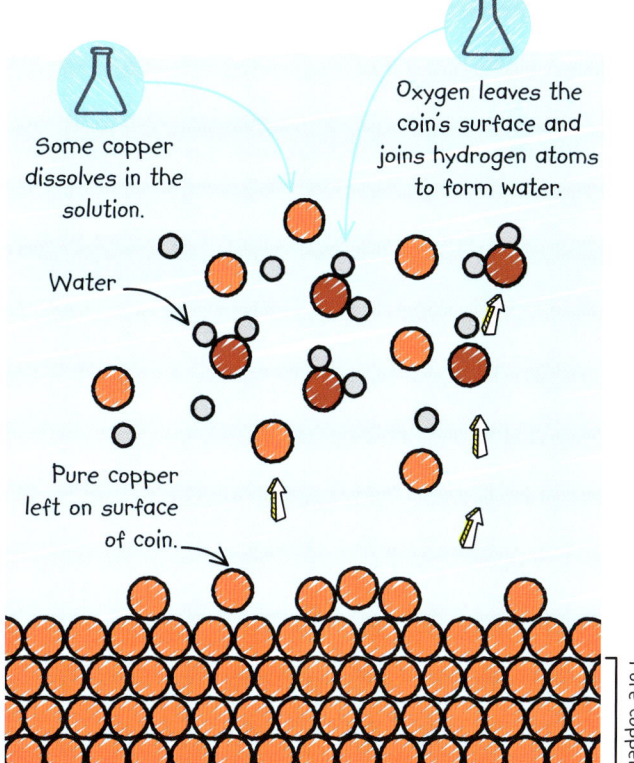

Some copper dissolves in the solution.

Oxygen leaves the coin's surface and joins hydrogen atoms to form water.

Water

Pure copper left on surface of coin.

Pure copper

LATER STAGE OF THE REACTION

EXPERIMENT 2 - COPPER PLATING

1 For the next experiment, use the vinegar and salt solution from the first experiment. It contains the copper from the coins. Put the steel nail into the solution.

The nail is now copper-plated.

2 After about 20 minutes, your nail should have changed colour. The copper from the coins now coats the nail.

HOW IT WORKS

Steel is a mixture of several different elements, but is made mostly of iron. Some of the iron atoms from the nail's surface dissolve in the solution – and some of the copper atoms from the solution (which originally came from the surface of the coins) attach to the nail instead. That's why the nail turns copper-coloured.

1. Copper atoms from the solution attach to the nail's surface.

2. Some iron breaks away and dissolves in the solution.

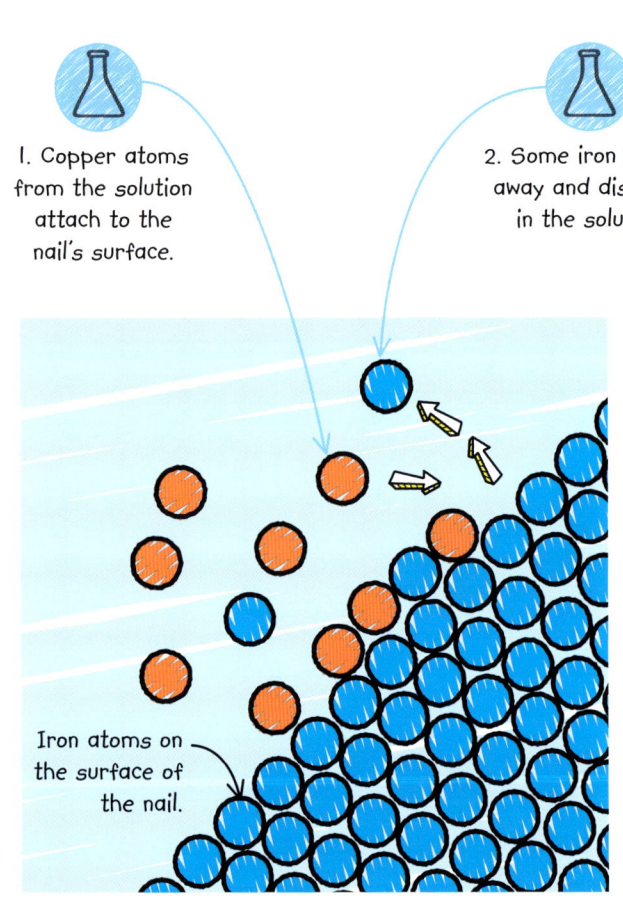

Iron atoms on the surface of the nail.

TEST AND TWEAK

Why not study in more detail the reactions between vinegar, salt, and copper. You can do this by letting the reactions go on for longer. Here are some of the stages of the reaction you might see.

1 Pour 50 ml (2 fl oz) of vinegar into the jug. Add half a teaspoon of salt. Stir until the salt is dissolved.

2 Put a coin in a jar and add the solution. Put the lid on. Leave it for a few days, opening daily to let in air.

EXPERIMENT 3 - MAKING COPPER CHLORIDE

1 Place a folded piece of paper towel into the plastic bowl. Soak the paper towel with vinegar.

2 Put one of the copper coins into the bowl with the vinegar-soaked kitchen towel.

3 Spoon salt on top of the coin, until the coin is completely covered.

The copper reacts with chlorine atoms from salt to form copper (II) chloride, which is blue-green.

4 Leave it for about an hour or two. You should see a green coating of copper chloride. Wash your hands after touching the coin.

HOW IT WORKS

When you leave a copper coin covered with salt and vinegar for a long time, a chemical reaction takes place. Salt is made of the elements sodium and chlorine. The chlorine reacts with the copper, producing a chemical compound called basic copper (II) chloride. This compound has a bright blue-green colour.

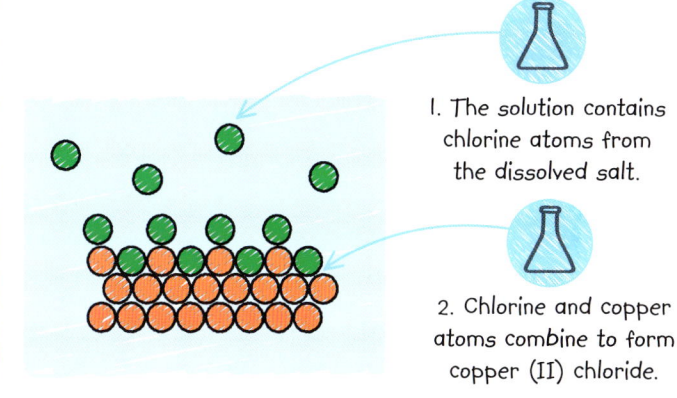

1. The solution contains chlorine atoms from the dissolved salt.

2. Chlorine and copper atoms combine to form copper (II) chloride.

Copper atoms join with chlorine atoms to form green copper (II) chloride.

As time passes and more air is let in, the solution turns brown.

Finally, the solution goes clear, but sooty deposits of copper (II) oxide can be seen in it.

SHAPES AND STRUCTURES

A building or a bridge needs to be strong enough to stay standing, and it also needs to hold up anything placed in or on top of it. But what makes a structure strong enough to support loads? It's all down to what it is made from, what shape it is, and how it's built. In this chapter, you'll be building some surprisingly strong structures using paper, sand, and drinking straws. You'll even build tall towers from spaghetti and marshmallows!

SPAGHETTI TOWER

For a tower to stay standing, it must be strong enough not to buckle under its own weight, and it must have a stable base. This tower is strong because of the shapes used in its construction: triangles. And its wide base stops it toppling over.

The tower must stand straight or the force of gravity will tip it over.

Triangles are strong because they can't twist out of shape.

The main part of the tower is made of two large cubes stacked together.

The base carries all the tower's weight.

HOW TO MAKE A
SPAGHETTI TOWER

All you need to build this tower is spaghetti, marshmallows, and willpower! The sticky marshmallows hold the ends of the spaghetti in place, and the spaghetti forms a sturdy framework. If you like marshmallows, you'll need willpower to stop yourself eating your building materials!

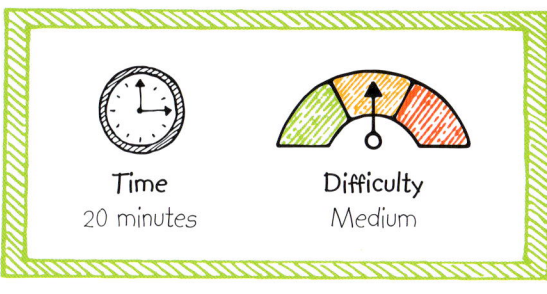

Time
20 minutes

Difficulty
Medium

STEM YOU WILL USE
• ENGINEERING: A structure will topple over if its centre of gravity is not supported.
• MATHS: A pyramid is a 3D shape with a base, triangular faces, and a pointed top.

WHAT YOU NEED

Spaghetti

Marshmallows

1 Begin by making a square. To see why a square isn't a strong shape, push it gently from one side. It leans easily, becoming a parallelogram.

Make sure the bottom marshmallows sit on their flat bases.

2 Make a cube. Try twisting it gently. Because it's made of squares, you'll find it leans very easily and isn't stable.

Sliding the marshmallows inwards helps to make the cube stronger.

3 To make the cube stronger, you'll need to add diagonal pieces. To fit them, first make the cube smaller by sliding the marshmallows inwards so the spaghetti strands poke out the other side.

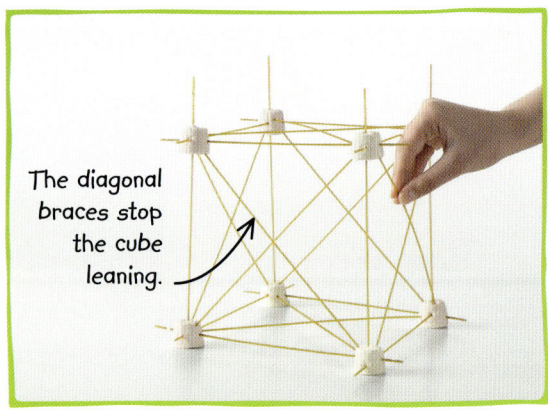

The diagonal braces stop the cube leaning.

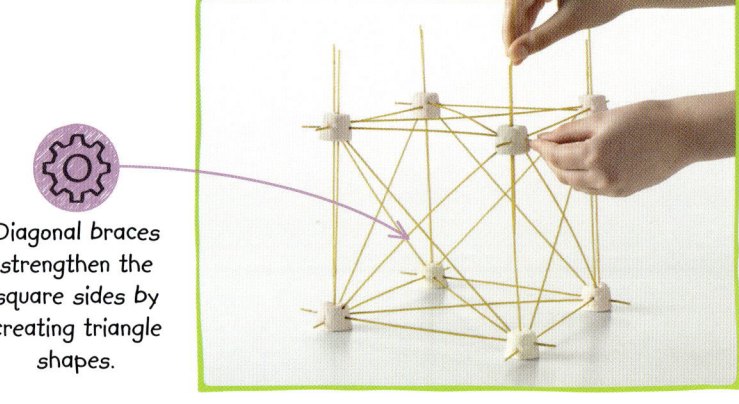

Diagonal braces strengthen the square sides by creating triangle shapes.

4 Add the diagonal pieces, called braces, across each face from corner to corner.

5 Strengthen the vertical edges by feeding a second piece of spaghetti down through the marshmallows at the top corners.

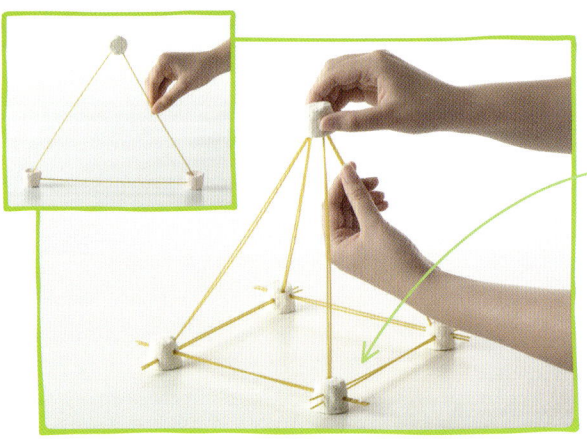

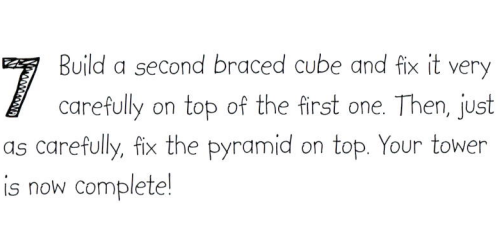

Make sure the square base of the roof is the same size as the top of your cube.

6 Make the roof, starting with a triangle. You'll notice this is stronger than a square as it doesn't lean. Add more spaghetti and marshmallows to form a pyramid with a square base.

7 Build a second braced cube and fix it very carefully on top of the first one. Then, just as carefully, fix the pyramid on top. Your tower is now complete!

Push the base of the top cube onto the bits of spaghetti sticking up from the first one.

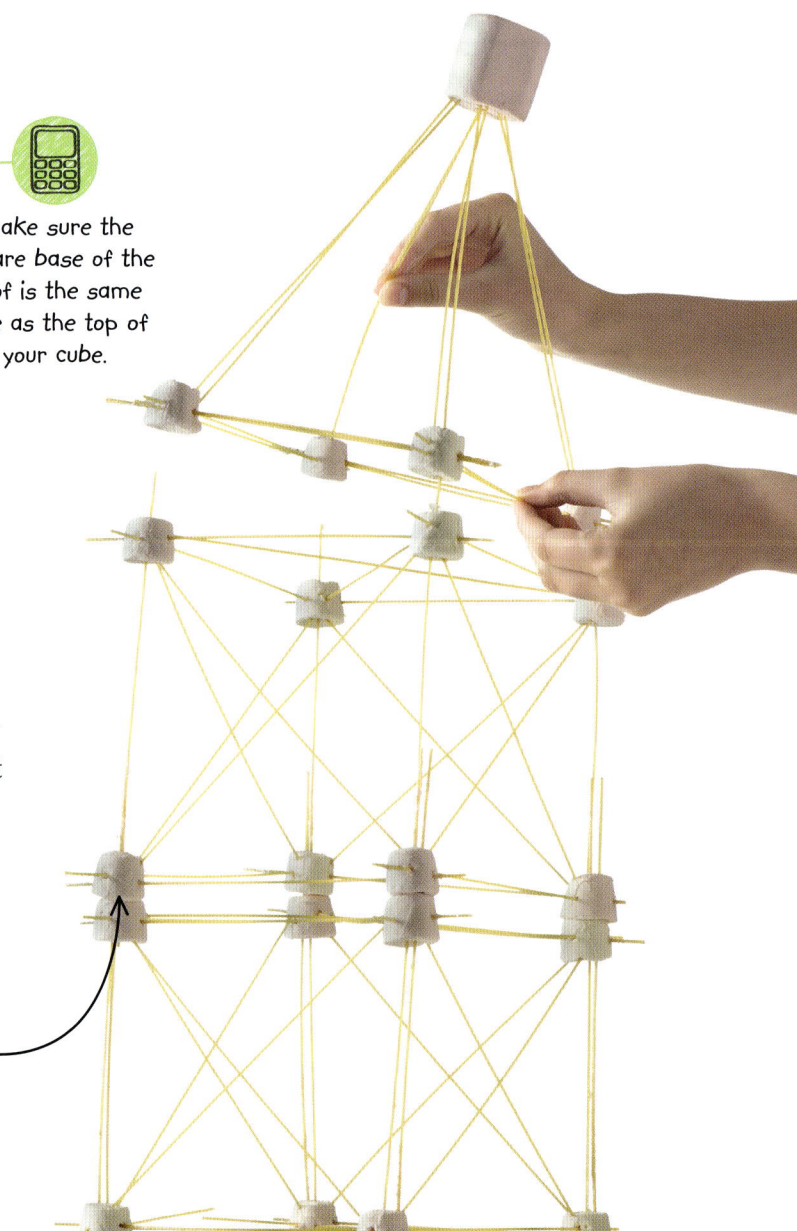

TAKE IT FURTHER

Now you've mastered the art of building spaghetti towers, why not try different designs? You could try building a tower that is one big pyramid. You'll need to plan the shapes carefully to make sure they all fit together. You might want to try to make your tower much taller. Can you make a tower that stands taller than you? The spaghetti pieces bend less if they're shorter – can you make a taller tower by using shorter pieces of spaghetti? If you want your tower to be stable, you'll need to make the base wide and the top the same width or narrower.

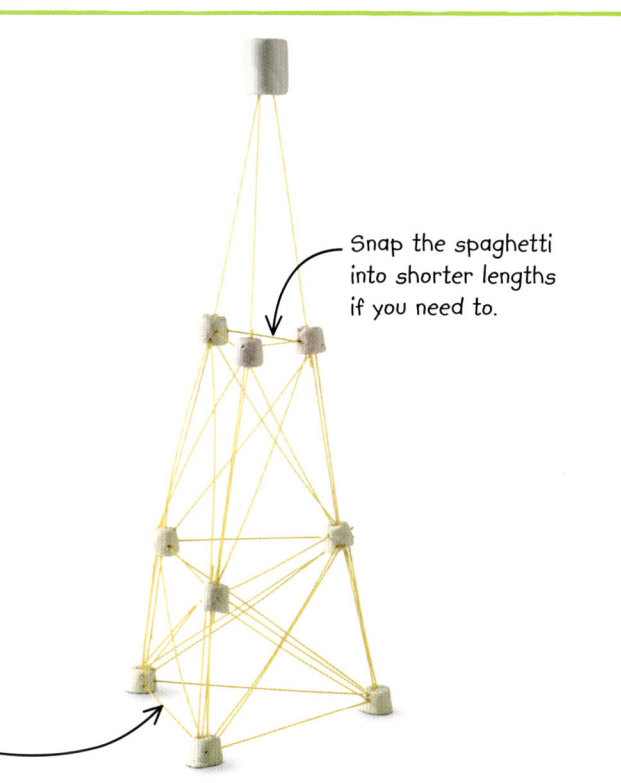

Snap the spaghetti into shorter lengths if you need to.

Try building a tower with a triangular base instead of a square base.

HOW IT WORKS

Triangles are the key to the strength of your tower. Unlike a square, which can lean over and turn into a parallelogram when pushed, a triangle can't change shape and so remains upright and rigid. The base of your tower must also be wide. For every object, there is an imaginary point called the centre of gravity. Its position is determined by the distribution of mass within the object. For the tower, it is closer to the base. Objects are stable if the centre of gravity is above the base. If an object leans, so that its centre of gravity is not above the base, it will topple.

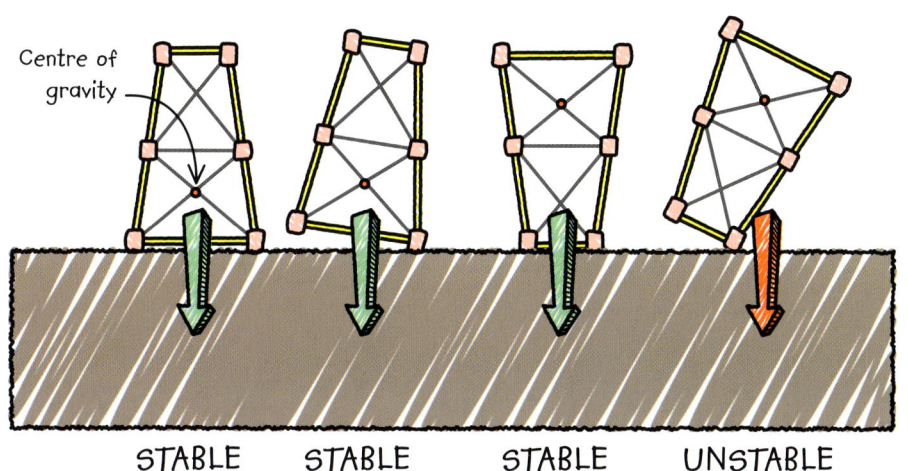

Centre of gravity

STABLE   STABLE   STABLE   UNSTABLE

REAL WORLD: ENGINEERING
TOKYO SKYTREE

With a height of 634 m (2,080 ft), the Tokyo Skytree in Japan is the tallest tower in the world. It's made of steel tubes arranged as strong triangles, and its base is much wider than its top.

NEWSPAPER STOOL

A single sheet of newspaper is very flimsy. It crumples and folds easily, and you would probably never think of using it to make anything really strong. But use lots of sheets of newspaper together, in just the right way, and you can make a stool so strong that it can support your weight!

Duct tape holds all the rolls together to make the stool.

Individual sheets of newspaper are not very strong at all.

Each roll of newspaper is held together by sticky tape.

HOW TO MAKE A
NEWSPAPER STOOL

You'll have to collect lots of newspaper to do this activity. If you want to make your stool really strong, you'll need to roll the newspaper very tightly. You might want to ask a friend to help, so that one of you can roll, and the other can stick. If you ever want to dismantle your stool, remove the tape first – then you can recycle the newspaper.

Time
45 minutes

Difficulty
Easy

WHAT YOU NEED

Sticky tape

Duct tape

Scissors

Lots of newspapers

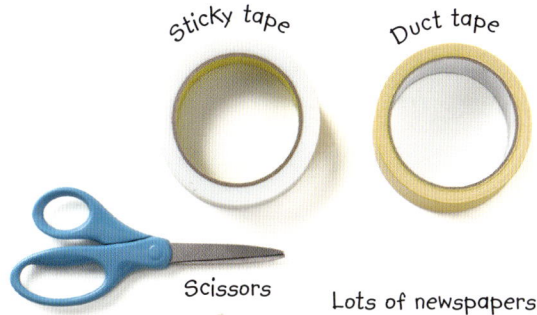

1 Roll up about 20 sheets of newspaper together lengthways. Roll it as tightly as you can. The roll should feel strong and rigid.

Try to keep the newspaper rolled as tightly as possible.

2 Wrap sticky tape around the roll, close to each end. You might want to ask someone to help you, as you need at least one hand to hold the roll. Repeat until you have 25 rolls.

3 Take another pile of about 20 sheets and cut it in half. You will use the two halves to make two short rolls.

4 Roll up and tape each half at the ends as before, so you finish with two rolls half the length of the others.

This duct tape is formed of three layers made from glue, fabric mesh, and plastic to make it water-resistant.

5 Line up eight long rolls with one short roll in the middle. Wrap duct tape around them, to secure them all together. Repeat this step with another eight long rolls and one short roll.

You now have two panels that can interlock with one another.

6 Slot one set of paper rolls into the other, using the gaps left by the short rolls in the middle of each. You should now have an X-shape that stands up on its own.

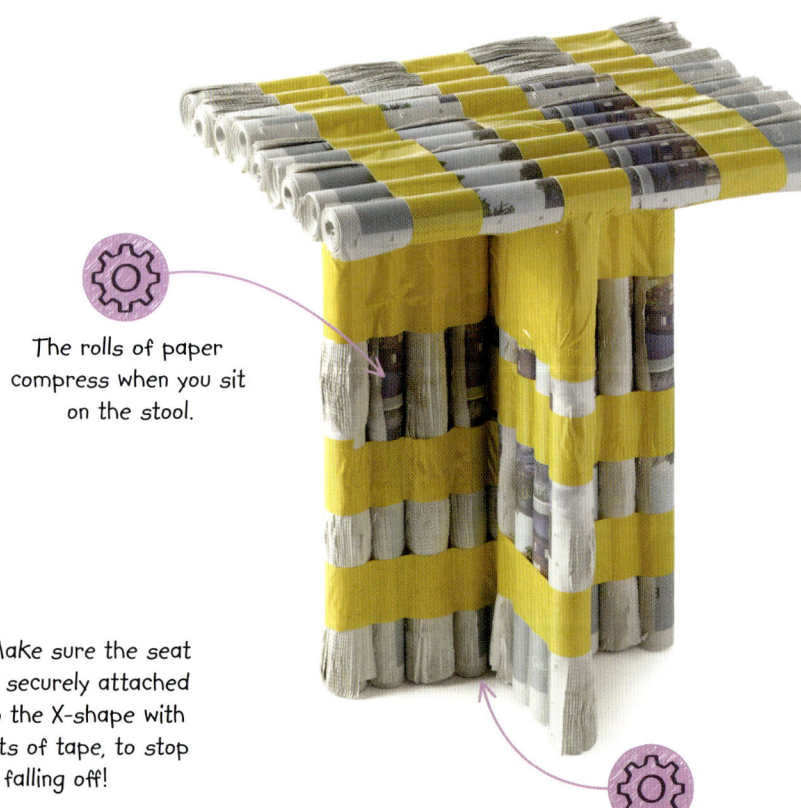

The rolls of paper compress when you sit on the stool.

The paper rolls are arranged as a cross, giving the stool stability.

Make sure the seat is securely attached to the X-shape with lots of tape, to stop it falling off!

7 To make your seat, line up the remaining nine long rolls of newspaper and secure them together with duct tape. Stand your X-shape on top of your joined rolls and use duct tape to attach it.

8 Stand your stool the right way up, so that the row of nine rolls is on top. Your stool is finished. Now, go ahead and sit on it!

HOW IT WORKS

The rolls of paper you used to make the stool are strong in two ways. Firstly, a cylinder has no corners, and so no point is weaker than any other, making it a very strong shape. Secondly, by rolling the paper rolls very tightly, you are increasing the density of the cylinders – packing more matter (stuff) into the same volume (space). If you made the rolls looser so they were much less dense, they wouldn't be as strong.

The force to support your weight comes from the molecules that make up the paper. When you sit on the stool, you squash, or compress, the paper slightly. The molecules of which the paper is made are pushed a tiny bit closer together, and that produces an equal force that pushes in the opposite direction – as if there are springs between them.

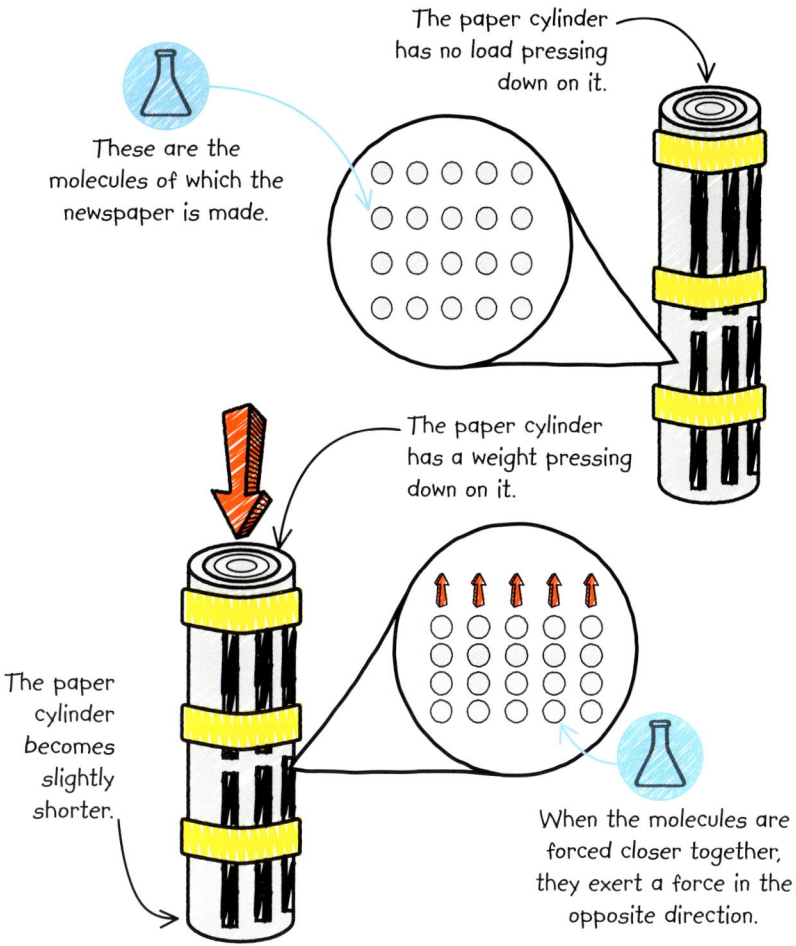

These are the molecules of which the newspaper is made.

The paper cylinder has no load pressing down on it.

The paper cylinder has a weight pressing down on it.

The paper cylinder becomes slightly shorter.

When the molecules are forced closer together, they exert a force in the opposite direction.

REAL WORLD: ENGINEERING
COMPRESSED COLUMNS

You can *see* cylindrical shapes like the paper rolls you used to make your stool in the columns that support *big* buildings. Just like your paper rolls, these columns are very strong in compression – even stronger than the paper, as they are made of dense stone.

REAL WORLD: SCIENCE
HOLLOW BONES

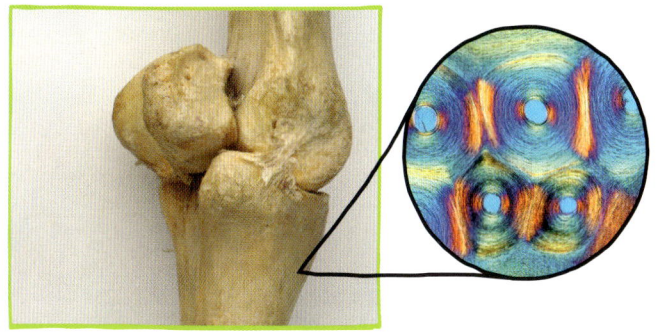

The long bones in your legs that support your weight work in a similar way to the paper rolls in your stool. They are hollow in the middle (to allow for the marrow, where *blood cells* are made), but very dense and strong around the outside. The dense part of the bone is made up of many small tubes (osteons), each one very weak. Just as *several* rolls of paper bound together can support a weight, so the clusters of osteons make the bone very strong.

SUSPENSION BRIDGE

Engineers build huge suspension bridges from concrete and steel cables. The concrete towers support the cables, and the cables support the road, which can carry hundreds of cars and lorries at a time. The best way to understand how these forces work together to make a strong and stable structure is to build your own model suspension bridge. In this activity, you can do just that by using bundles of drinking straws instead of concrete, and string instead of steel cables.

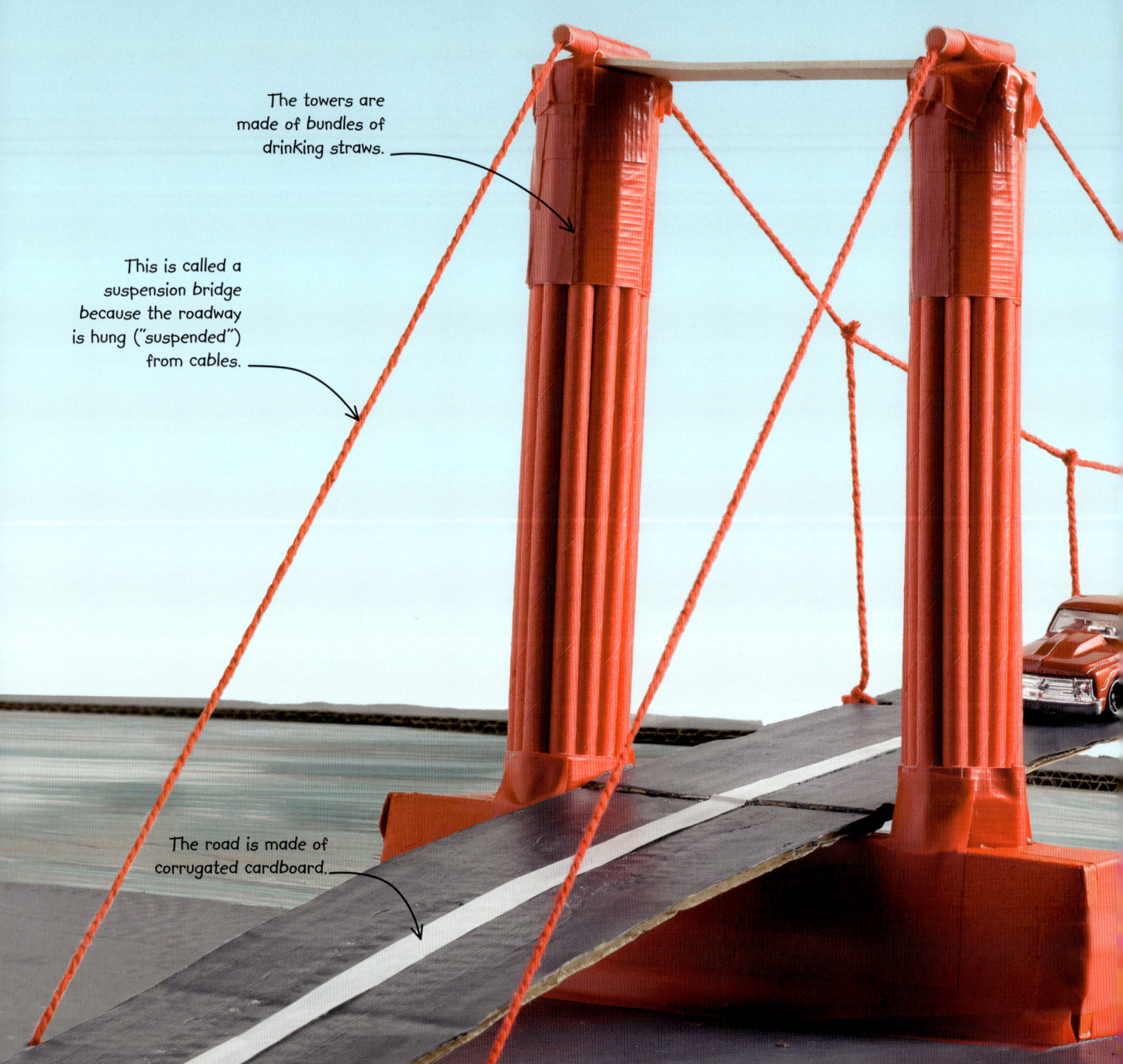

The towers are made of bundles of drinking straws.

This is called a suspension bridge because the roadway is hung ("suspended") from cables.

The road is made of corrugated cardboard.

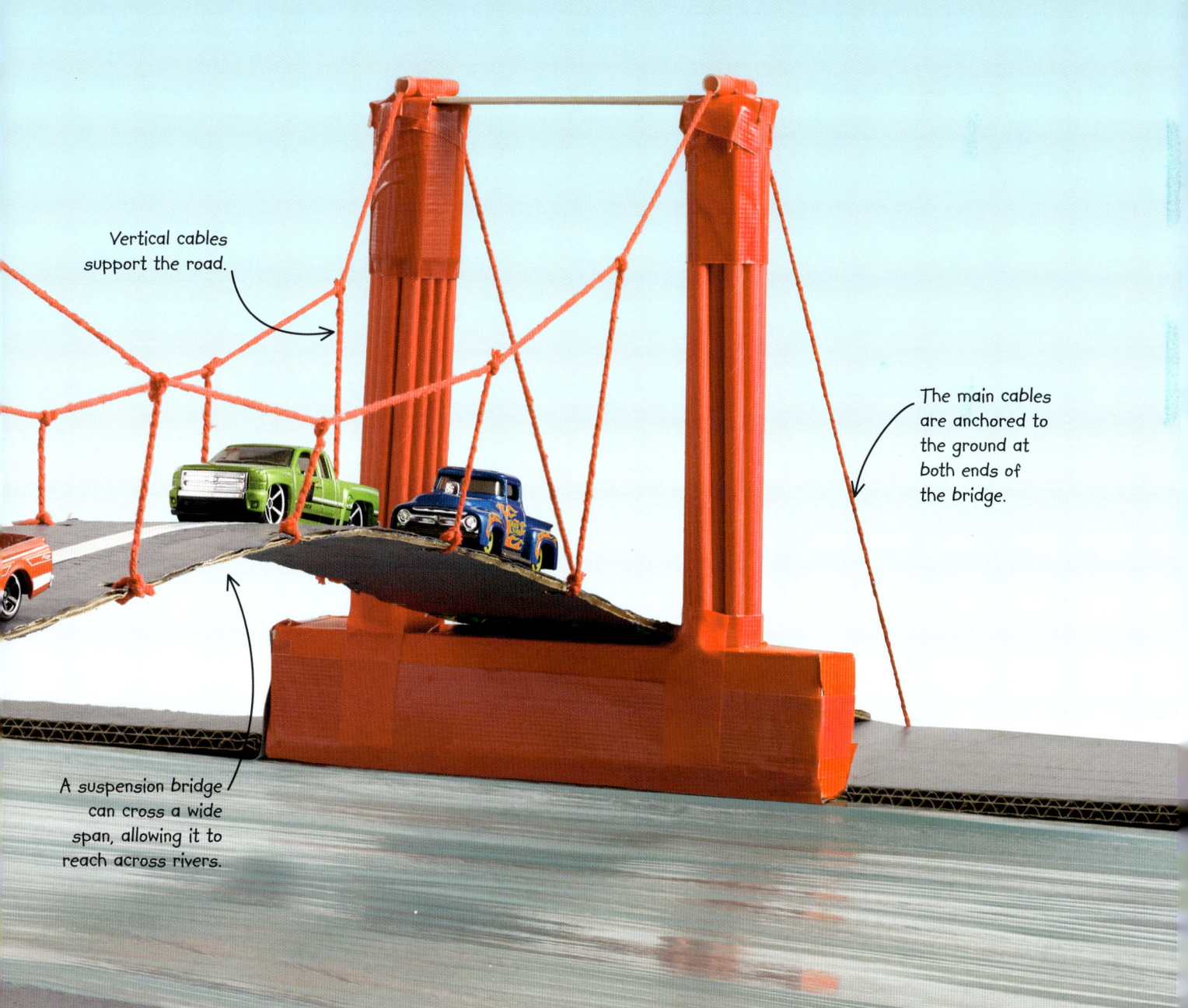

Vertical cables support the road.

The main cables are anchored to the ground at both ends of the bridge.

A suspension bridge can cross a wide span, allowing it to reach across rivers.

HOW TO MAKE A
SUSPENSION BRIDGE

The bases of the bridge's towers are made from toothpaste tube boxes. If you can't find any, you can make boxes the right size from cardboard. The towers are made from bundles of drinking straws. We used 15 in each bundle – if your straws are wider or narrower than ours, use fewer or more straws.

Time
I hour

Difficulty
Hard

WHAT YOU NEED

Double-sided tape

Strong tape

String

Rubber band

Hole punch

Scissors

Pencil

Grey paint

Four lollipop sticks

Black felt-tip pen

Paintbrush

White paint

Small pebbles

Ruler

Two toothpaste boxes, each 5 cm x 5 cm x 20 cm (2 in x 2 in x 8 in)

Lots of corrugated cardboard

61 paper straws

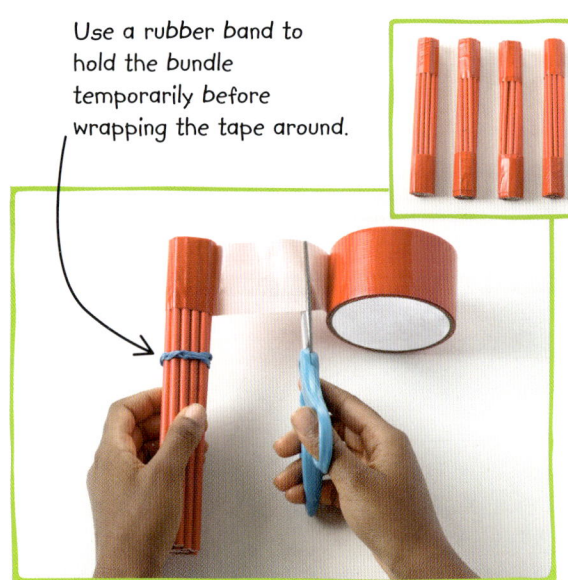

Use a rubber band to hold the bundle temporarily before wrapping the tape around.

1 Make a bundle of fifteen straws. Now wrap a piece of wide, strong tape around each end. Repeat three times, to make a total of four towers.

2 Firmly tape two lollipop sticks side-by-side across the tops of two of your towers. Repeat with the other two towers.

3 Cut four 2½ cm (1 in) pieces from the last straw. These will hold the bridge's two main cables in place.

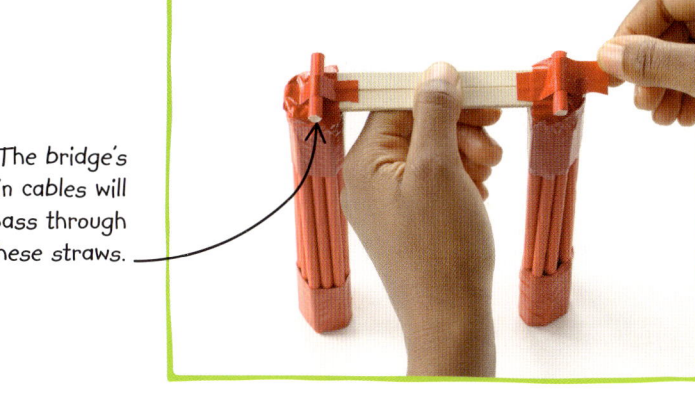

The bridge's main cables will pass through these straws.

4 Tape each piece of straw to the top of a tower at a 90° angle to the lollipop sticks.

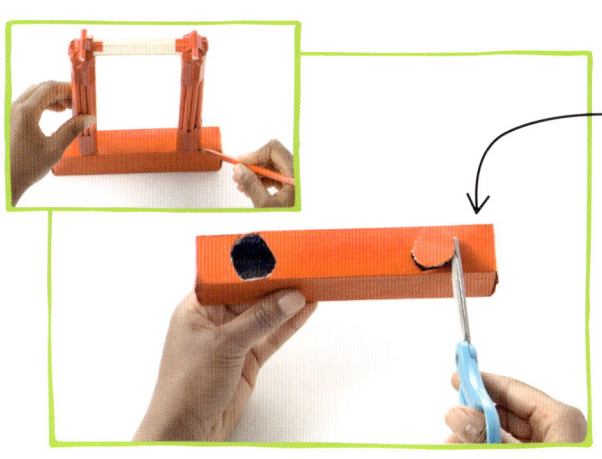

If you like, paint the toothpaste boxes first.

5 Draw around the bases of your towers on one side of each toothpaste box. Cut around the lines and push the towers into the holes.

The pebbles are ballast – their weight makes the bridge stable.

6 Fill both boxes with the pebbles. You may need to adjust the position of the towers to fit the pebbles around them. Close the boxes and secure them with strong tape.

7 Cut out two 20 cm (8 in) wide cardboard squares and, if you want, paint the tops grey. These will form the base your bridge stands on.

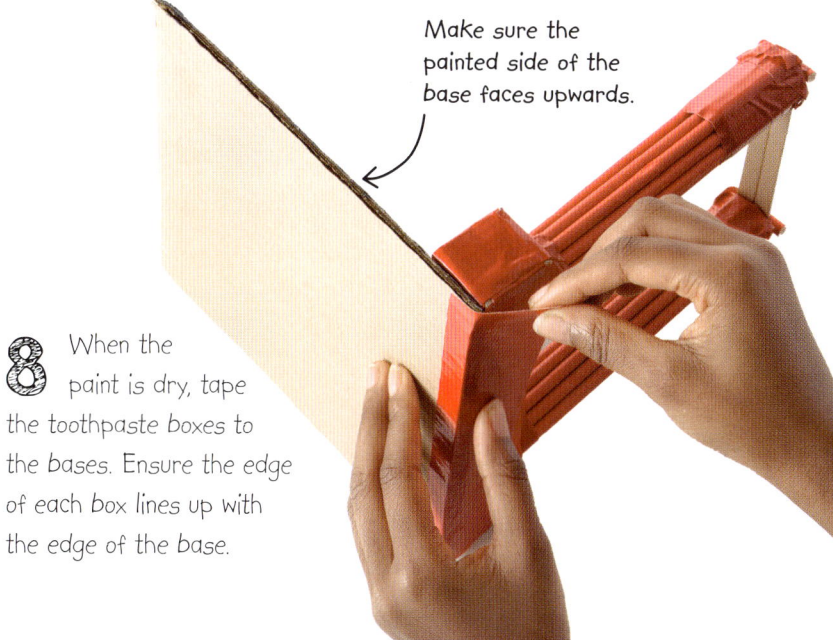

Make sure the painted side of the base faces upwards.

8 When the paint is dry, tape the toothpaste boxes to the bases. Ensure the edge of each box lines up with the edge of the base.

If you like, paint a white line in the centre of the road.

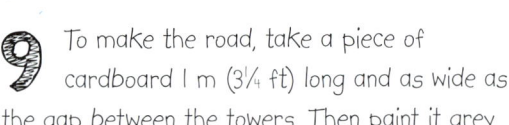

9 To make the road, take a piece of cardboard 1 m (3¼ ft) long and as wide as the gap between the towers. Then paint it grey.

10 Add a piece of double-sided tape between the two towers on each base. Remove the protective strip so the tape is sticky.

Tape the road down here.

11 Press the road onto the double-sided tape at both ends, as shown here, and then tape the end of the road to the edge of the base.

A real bridge's roadway might crack rather than sag, if it was made of concrete, which can be brittle.

12 If you put objects on the bridge now, it will sag. That's because the bridge still needs cables to support heavy weights.

13 Make pen marks 10 cm (4 in) apart along both sides of the road to mark the points where cables will attach.

Take care not to punch holes through the edge of the road.

14 Use a hole punch to make a hole at each mark. Make sure you push the hole punch well onto the road so that the hole isn't too near the edge.

15 Now make your cables. Cut two long pieces of string, each 1½ m (5 ft) in length. Then cut 10 short pieces of string, each 15 cm (6 in) long.

16 Thread the long pieces of string through the short straws on top of both ends of the bridge. You should then have two parallel "cables".

The long pieces of string will be the bridge's main cables. In a real bridge, the cables are made from steel.

17 Cut short slits in the base at both ends of the bridge, and wedge the ends of the main cables into the slits. Don't pull the string tight – it should hang in the middle of the bridge.

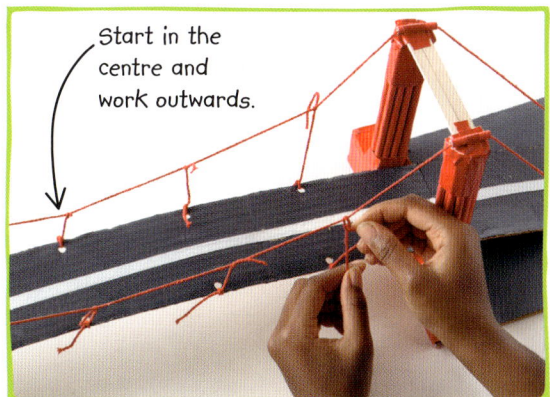

Start in the centre and work outwards.

18 Tie the short pieces of string between the holes in the road and the main cables. Starting from the middle, they should be about 3 cm (1¼ in), 5 cm (2 in), and 7 cm (2¾ in) long.

19 Pull the main cables through the slits in the bases so that they're taut. Then secure the ends of the bridge to a table or board with tape. Your suspension bridge is now complete!

20 Try placing objects on the road to see how well the bridge supports their weight. When the bridge is carrying a heavy load, touch the vertical strings and the main cables to see if you can feel them become more taut.

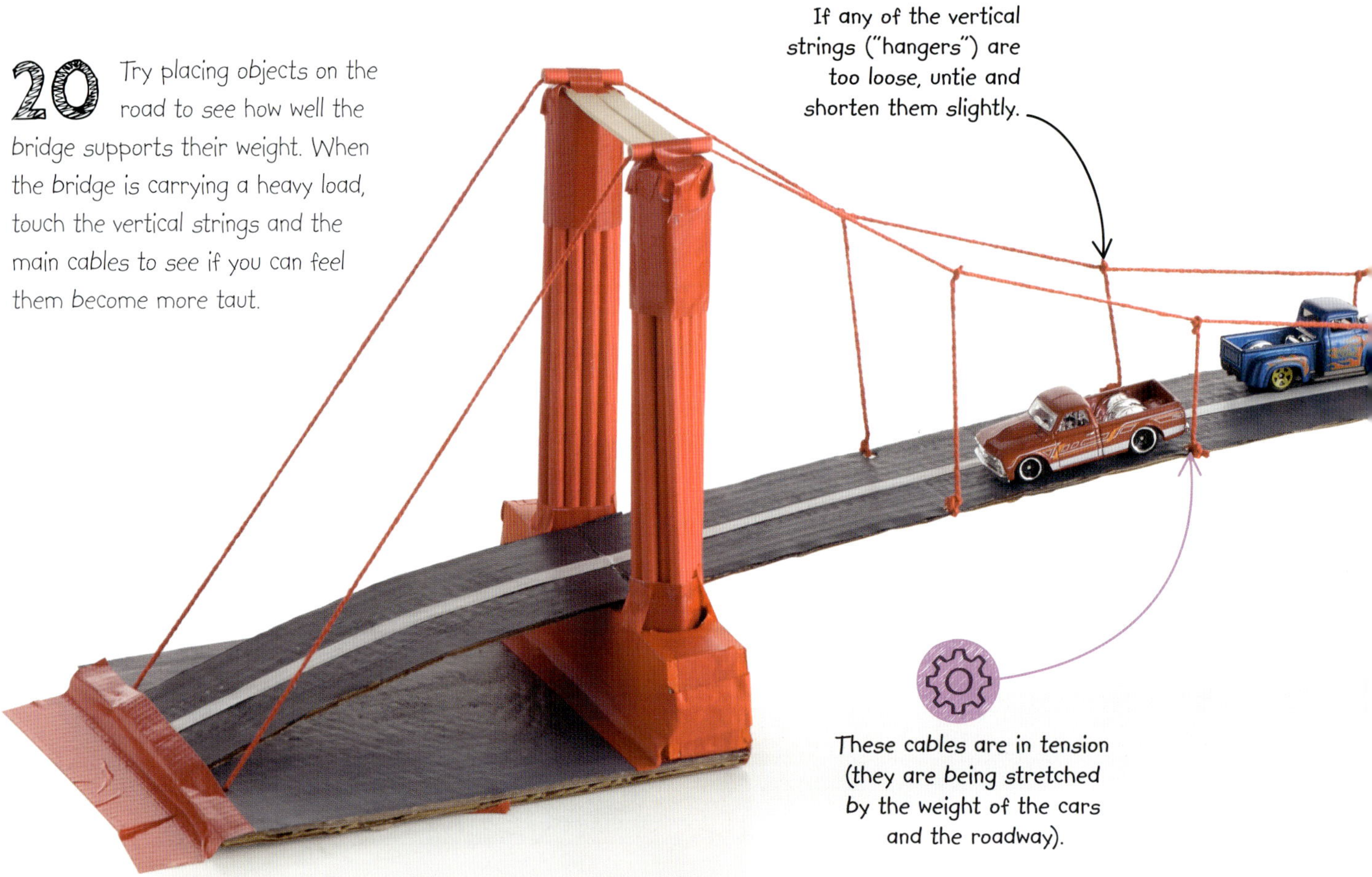

If any of the vertical strings ("hangers") are too loose, untie and shorten them slightly.

These cables are in tension (they are being stretched by the weight of the cars and the roadway).

HOW IT WORKS

If you put objects on your bridge when it isn't supported by cables, the load will make the road sag. A real bridge without cables would break apart under a heavy load. The bridge must be able to push upwards on the load to support it. This upward force is supplied by the tension in the vertical strings – and they are supported at the top by tension in the main cables. The main cables are supported by the towers, which are themselves supported by the ground.

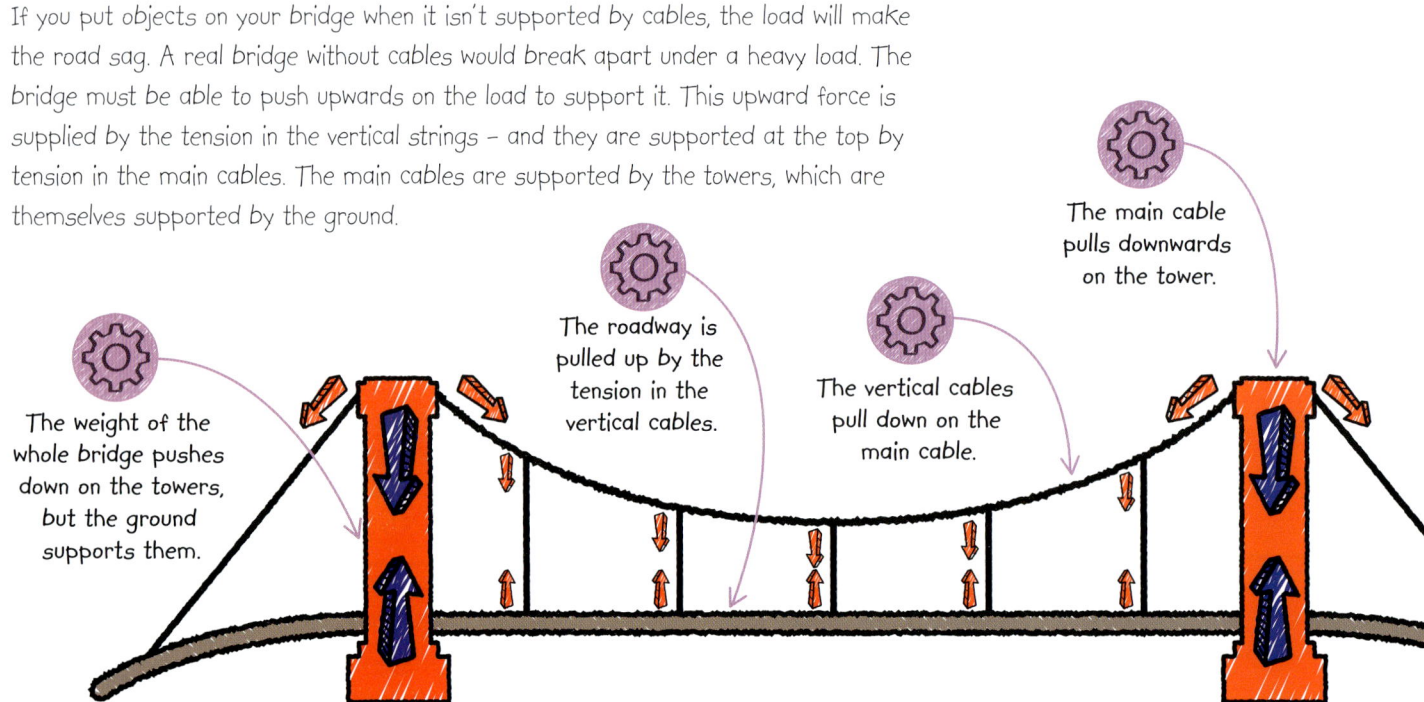

The main cable pulls downwards on the tower.

The roadway is pulled up by the tension in the vertical cables.

The vertical cables pull down on the main cable.

The weight of the whole bridge pushes down on the towers, but the ground supports them.

The roadway should arch in the middle slightly when the bridge is finished.

Suspension bridges can achieve a wider span than any other kind of bridge.

The main cables must be firmly anchored at each end of the bridge to maintain their tension.

TAKE IT FURTHER

Once you know how to build a model suspension bridge, why not try making a longer, taller, or wider one. Can larger bridges carry just as much weight or do they need more cables to support them? Do the main cables need to be anchored to the ground or would the bridge still work if they're only anchored to the road? What's the maximum weight your bridge can support? Load it until it collapses to find out!

REAL WORLD: ENGINEERING
GOLDEN GATE BRIDGE

Perhaps the best-known suspension bridge in the world is the Golden Gate Bridge just north of San Francisco, USA. Around 112,000 vehicles cross this bridge every day. The suspended roadway is 2.7 km (1⅔ miles) long and is held high above the sea by two main cables measuring 2.3 km (1⅖ miles) long.

REAL WORLD: TECHNOLOGY
CABLE SUPPORTED ROOF

BC Place sports stadium in Vancouver, Canada, has a fabric roof held up by 35 km (22 miles) of steel cables. The cables are supported by 36 steel towers, which do the same job as the towers in a suspension bridge. The fabric roof is strong enough to support 7,000 tonnes of snow and can retract to create an open-air stadium when the weather is good.

GEODESIC DOME

This impressive structure is known as a geodesic dome. It's easy to make, and though it is quite light and looks fragile, it is actually extremely sturdy because of its shape. Once you've made your dome, you can cover it with clear cellophane to make a small greenhouse.

Triangles make the structure sturdy and stable.

Cellophane traps heat from the Sun inside the greenhouse, making it warmer.

STEM YOU WILL USE
• SCIENCE: Greenhouses allow sunlight in, but do not allow much heat to escape.
• ENGINEERING: With struts (bars) connected in triangles, even a light structure can be strong.
• MATHS: Domes are normally hemispheres (half spheres).

The straws are the struts.

Succulent plants can survive in hot environments with little rain, such as the inside of a greenhouse.

Pipe cleaners, hidden by the tape, connect the struts.

HOW TO MAKE A
GEODESIC DOME

This geodesic dome is made of 65 struts, of two different lengths, joined together by connectors made from pipe cleaners. We've used two kinds of paper straw, to distinguish the long struts from the short struts, and two kinds of pipe cleaner: one for the feet connectors at the base of the dome, and one for the regular connectors. You don't have to use the same colours as we have.

Time
1 hour

Difficulty
Hard

WHAT YOU NEED

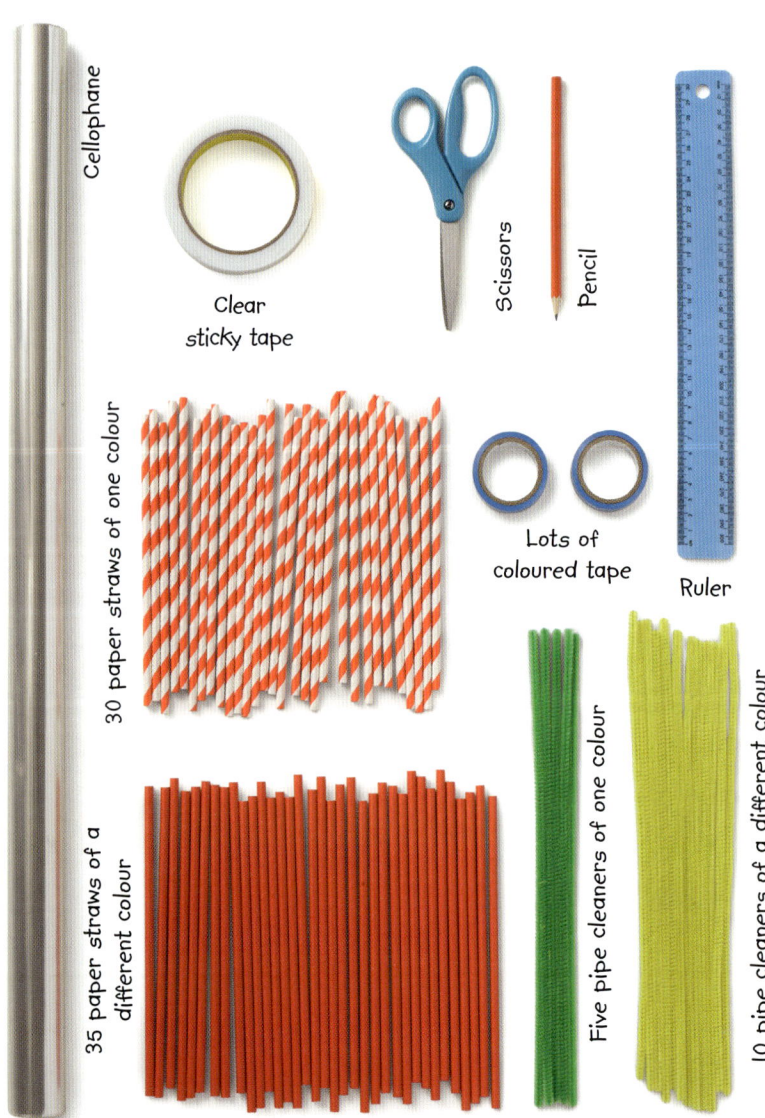

Cellophane

Clear sticky tape

Scissors

Pencil

Lots of coloured tape

Ruler

30 paper straws of one colour

35 paper straws of a different colour

Five pipe cleaners of one colour

10 pipe cleaners of a different colour

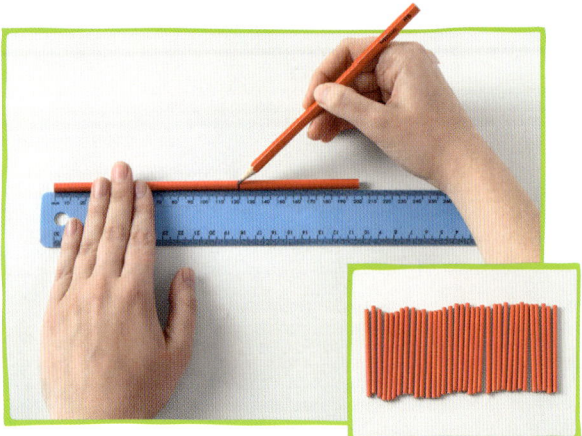

1 First, make the 35 long struts from straws. They should be 12 cm (5 in) long. For each one, draw a line first, then cut the straw at the line.

2 Now make 30 short struts from the other straws. These should be 11 cm (4½ in) long. Make sure you recycle the bits of straw you don't need from steps 1 and 2.

You should end up with 20 of one colour and 40 of the other.

3 Gather together five pipe cleaners of one colour and 10 of the other. Fold each one in half and cut, and then cut each half in half again.

4 Twist together pairs of pipe cleaners from the pile of 20, just like in the picture here, to make 10 "feet" for your dome.

A flat 10-sided shape is called a decagon.

5 To make the connectors, twist together three lengths of the other pipe cleaner. You'll need 12 of these connectors altogether.

6 Connect 10 of the long struts by using the pipe-cleaner feet. You'll end up with a 10-sided shape.

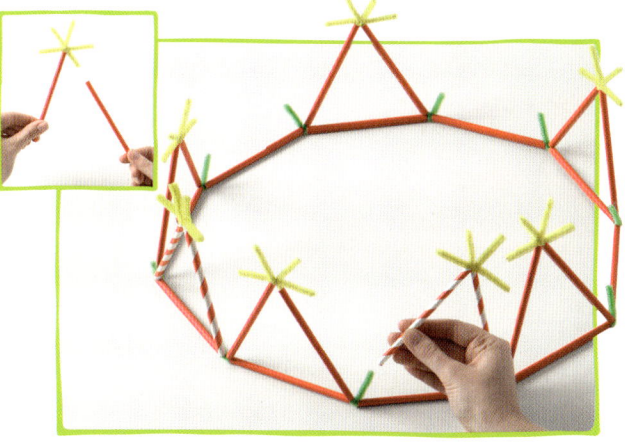

7 Use the connectors to begin building up the dome, forming the bottom layer of triangles. Alternate the long and the short struts, as shown.

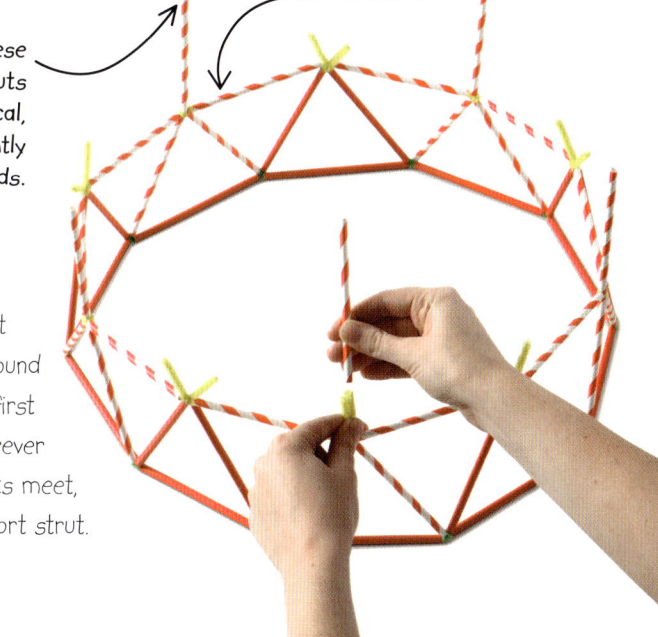

Short strut

Make these short struts nearly vertical, leaning slightly inwards.

8 Add short struts around the top of the first layer, then wherever four short struts meet, add another short strut.

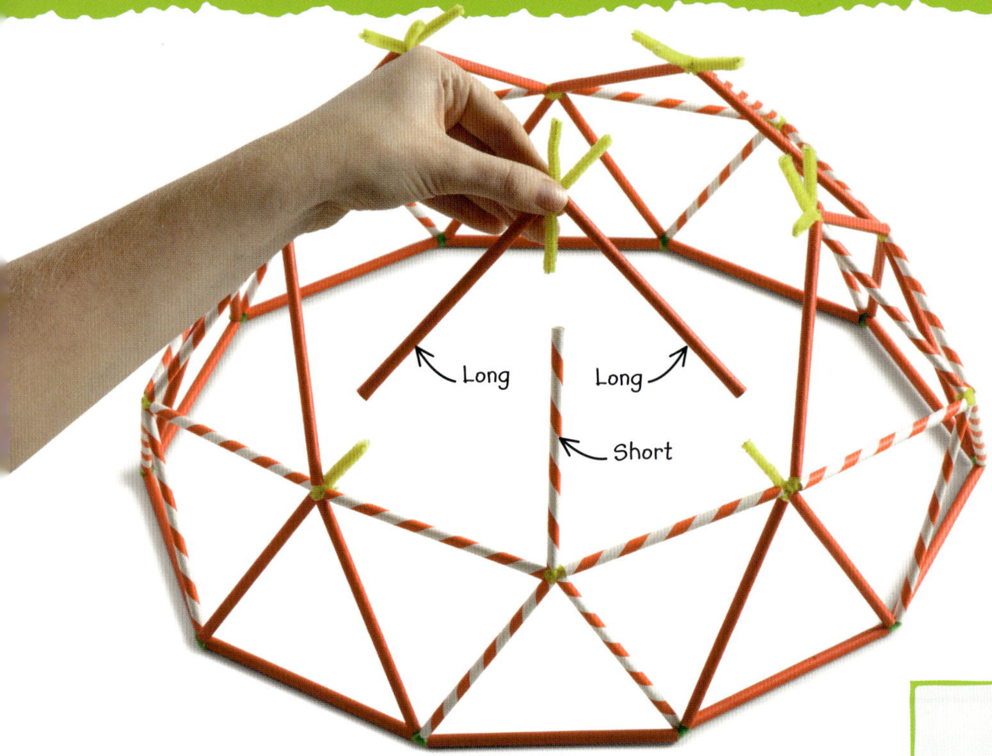

Long   Long

Short

9 Around each near-vertical short strut, use a connector to add two long struts, as shown.

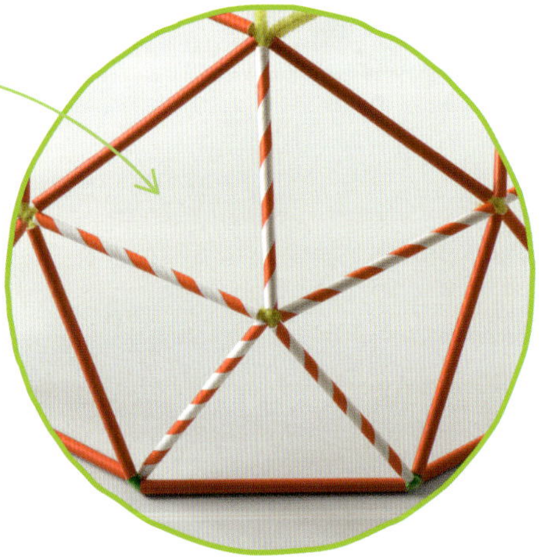

The short struts form five triangles inside each pentagon.

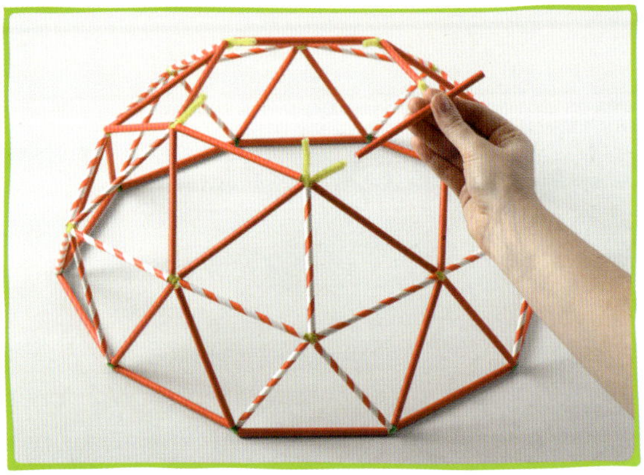

10 Connect five more of the long struts around the top of the middle layer. These will form a pentagon on top of the dome.

One strut will have two pipe cleaner legs tucked into it.

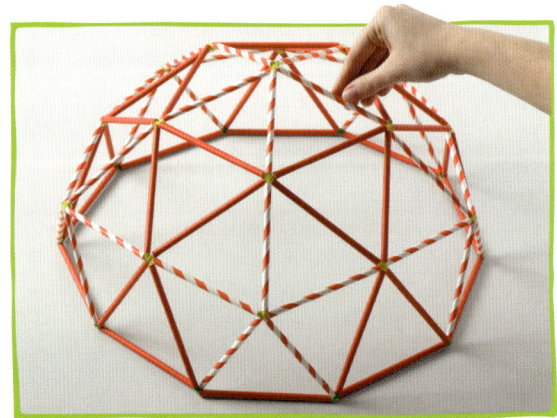

11 To complete the top of the dome, join five short struts with a connector. Tuck the spare sixth leg of the connector into one of the struts.

12 Join the five remaining short struts to the spare connectors at the top of the dome.

The cellophane covering will turn your geodesic dome into a small greenhouse ready for plants.

13 Wrap short lengths of coloured tape around all the joints to strengthen the structure.

14 Now cover your dome in cellophane and secure the pieces in place with clear sticky tape. Your geodesic dome is complete!

HOW IT WORKS

The geodesic design is very sturdy because it has the stability of the triangle shape as its building block. Triangles are strong because they don't distort when put under pressure. If pressure is put on one corner of a triangle, the other two corners distribute the force evenly. In your geodesic dome, the triangles are repeated, so any force on the building divides repeatedly at each intersection and spreads efficiently through the structure.

In a geodesic dome, the weight of the building is distributed efficiently across the whole structure.

The forces are reduced at each level of the dome.

The triangle shape divides the forces evenly at every intersection.

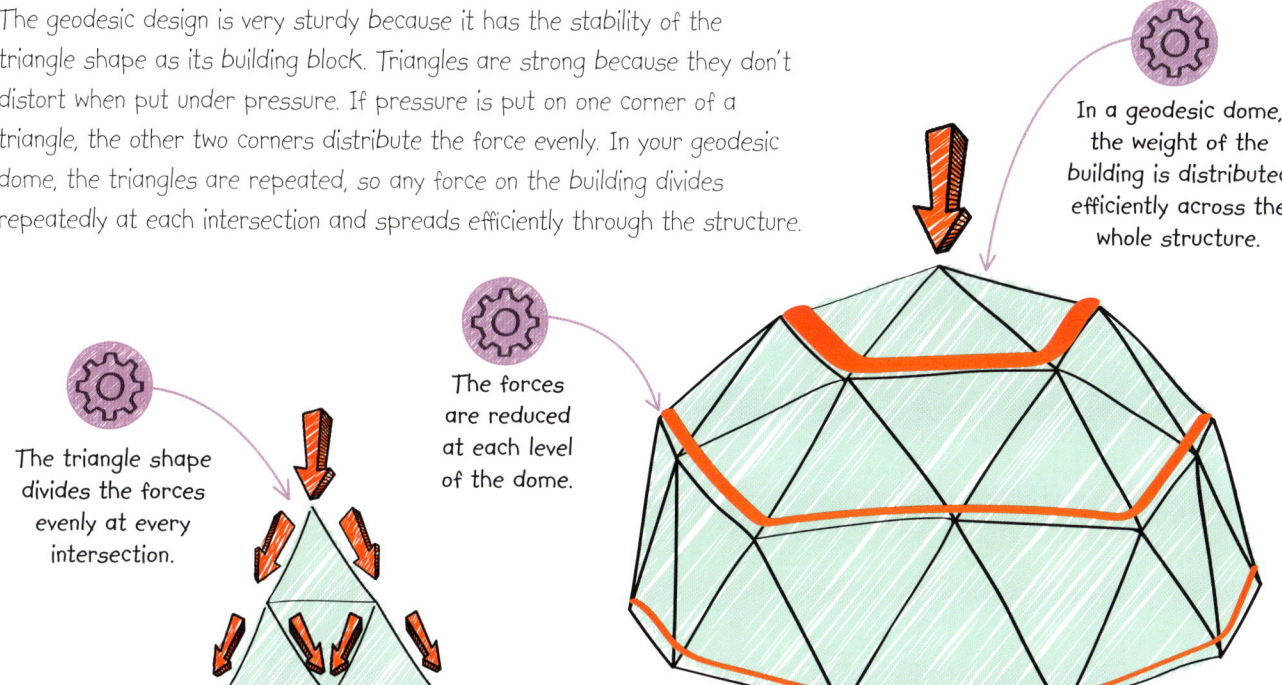

REAL WORLD: SCIENCE
BUCKMINSTERFULLERENE

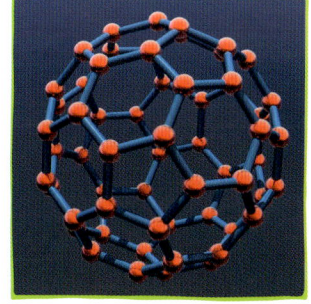

In 1986, scientists discovered a form of the element carbon whose atoms are arranged in a geodesic shape of pentagons and hexagons. This form of carbon was named buckminsterfullerene, after one of the most important designers of geodesic domes, the American architect Richard Buckminster Fuller.

PANTOGRAPH

Invented in 1603, the pantograph is a device that was once used to copy drawings and enlarge them at the same time. It's made of four rigid pieces that pivot around joints. When you move the pencil in the middle, the end of the right arm copies the movement of your hand but covers greater distances. The effect is quite spooky, as though an invisible hand is holding the second pen and copying you. Why not try this spooky drawing machine for yourself?

Joints made with paper fasteners allow the pantograph to flex and stretch.

The whole pantograph pivots around this point, which is fixed to the table with adhesive putty.

Move this pencil with your hand to draw a picture.

This pen draws an enlarged copy of your picture.

HOW TO MAKE A
PANTOGRAPH

Your pantograph is made of four rectangles of cardboard joined by paper fasteners. It's important that the cardboard can move freely at the joints. The pantograph works *best* if you draw a simple picture in a single, continuous line without lifting the pencil.

Time
30 minutes

Difficulty
Medium

WHAT YOU NEED

Felt-tip pens

Adhesive putty

Sticky tape

Pencil

Paintbrush

Two bottle caps

Scissors

Paper fasteners

Paint

Corrugated cardboard

Paper

Ruler

1 Use the pencil to draw three rectangles on the cardboard, each 50 cm (20 in) long by 5 cm (2 in) wide. Cut them out.

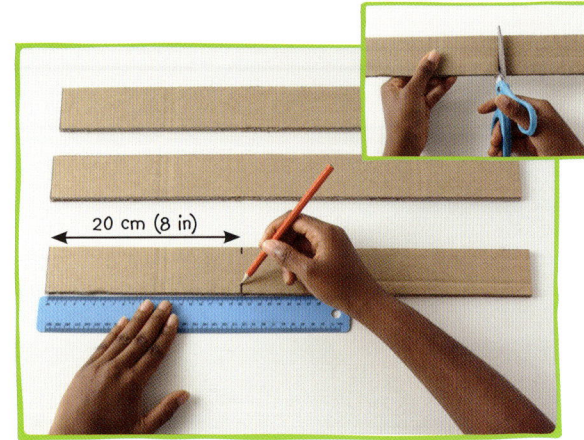

20 cm (8 in)

2 Use a pencil to mark a line on one of the rectangles, 20 cm (8 in) from the end. Cut across this line to make two rectangles, one 20 cm (8 in) long and the other 30 cm (12 in).

3 You now have all four pieces for your pantograph. If you like, paint them and allow the paint to dry.

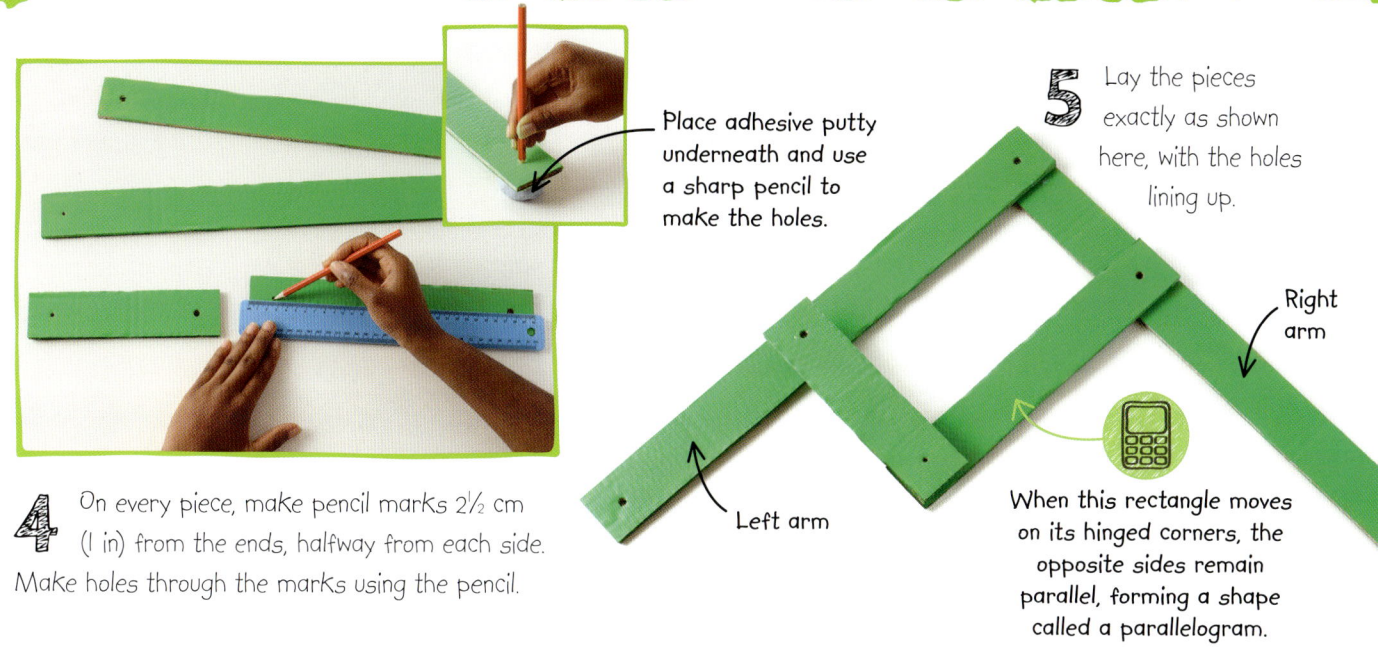

Place adhesive putty underneath and use a sharp pencil to make the holes.

4 On every piece, make pencil marks 2½ cm (1 in) from the ends, halfway from each side. Make holes through the marks using the pencil.

5 Lay the pieces exactly as shown here, with the holes lining up.

Right arm

Left arm

When this rectangle moves on its hinged corners, the opposite sides remain parallel, forming a shape called a parallelogram.

Joints are connections between moving parts in a machine.

Don't fasten this joint yet.

6 Poke paper fasteners through the two holes shown and fold the metal wings back. Don't put a fastener in the other holes yet.

7 Place a bottle cap on a lump of adhesive putty and use a sharp pencil to make a hole in it. Do the same to the other bottle cap.

Use a paper fastener to secure the bottle cap.

8 Use a paper fastener to fix a bottle cap to the end of the left arm. Push in a lump of adhesive putty. This will secure the arm to the table.

9 Fix the other bottle cap underneath the joint between the two arms. This will keep the cardboard parts raised above the table.

10 Wrap tape around the ends of the two short bits of cardboard and the free end of the right arm. This will stop the cardboard splitting when you push the pencil and pen through.

11 Push the pencil through the taped ends of the short bits of cardboard and leave it in place. Then make a hole through the end of the long arm and push the felt-tip pen through it.

12 To use the pantograph, draw a picture with the pencil and watch the felt-tip pen as it makes a larger copy. Try to do this in one continuous movement, without lifting the pencil or repeating lines. You can also use the pantograph to trace existing pictures and enlarge them.

The pantograph acts as a lever that magnifies movement rather than forces.

If the pantograph won't stay in place as you use it, hold the fixed end down with one hand.

HOW IT WORKS

The pantograph is an example of a mechanical linkage – a machine made of rigid pieces that are joined but can still move. A mechanical linkage changes one kind of movement into another. In this case, the movement you put in is magnified. In the centre of the pantograph is a parallelogram – a shape whose opposite sides are parallel (lined up in the same direction). The pencil and pen are mounted on parallel parts, so they trace out the same shape as they move. However, because the pen is on a longer arm, the shape it draws is magnified. The magnification equals the length of the pen's arm divided by the length of the pencil's arm (A ÷ B in the diagram).

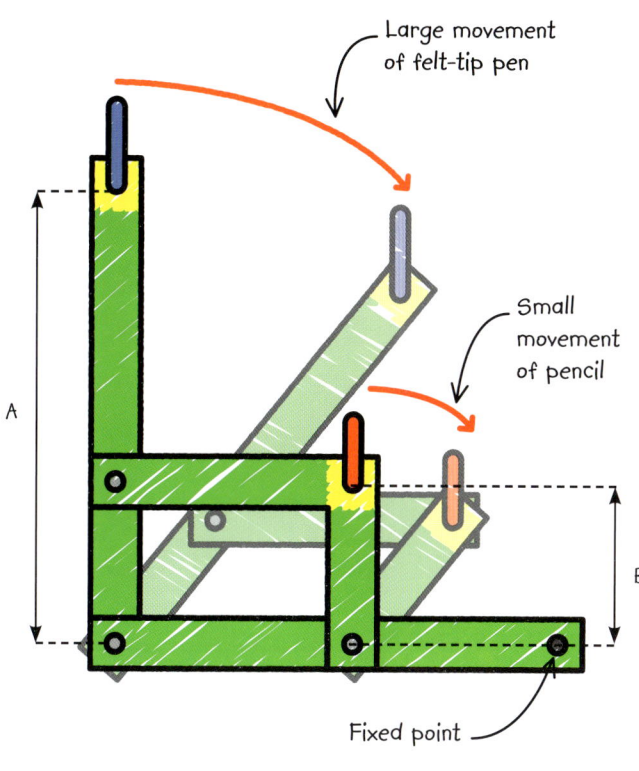

Large movement of felt-tip pen

Small movement of pencil

A

B

Fixed point

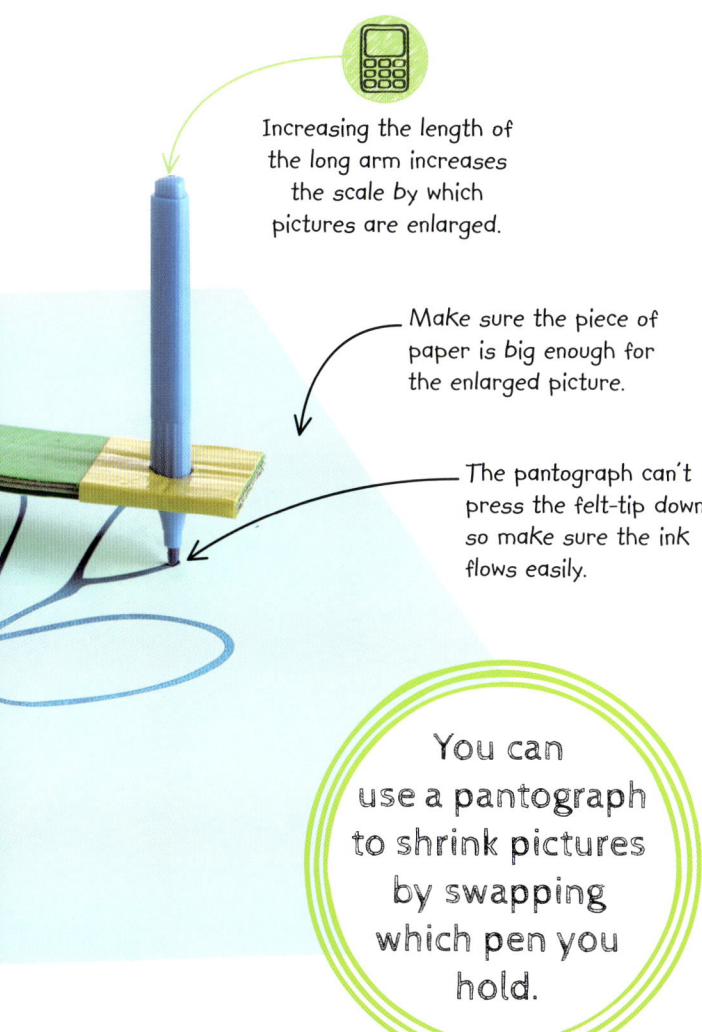

Increasing the length of the long arm increases the scale by which pictures are enlarged.

Make sure the piece of paper is big enough for the enlarged picture.

The pantograph can't press the felt-tip down, so make sure the ink flows easily.

You can use a pantograph to shrink pictures by swapping which pen you hold.

REAL WORLD: SCIENCE
UMBRELLA

Mechanical linkages are found in many different machines and devices, including something you probably have at home: an umbrella. When you push up on the sliding part inside an umbrella (the rider), the rods and pivots inside the umbrella magnify the movement of your hand to lift up the entire canopy, giving you shelter from the rain.

STURDY SANDCASTLE

If you've ever made a sandcastle, you'll know you have to use a bit of water to make the sand damp in order to bind the grains together. Try it with dry sand and you'll end up with a heap because there's nothing to stick the grains together. But even a castle made with damp sand will collapse under a little weight. With the help of a little science and engineering, however, this project reveals how you can make a super-strong sandcastle – one that might even be able to support your weight!

STEM YOU WILL USE
• SCIENCE: Water molecules are attracted to sand molecules, that's what makes them clump together easily.
• ENGINEERING: Reinforcing materials, such as bandages, can make a structure stable and support much heavier loads.

In wet sand, the water binds the sand grains, helping to hold the structure together.

There is a force caused by the books pressing downwards on the sandcastle.

The sandcastle is "reinforced", which means it has been strengthened with extra material.

Layers of bandages give this castle extra strength.

HOW TO BUILD A
STURDY
SANDCASTLE

There is a special ingredient that makes this sandcastle strong: strips of bandage. Apart from that, it's just like most sandcastles: you fill a bucket with damp sand, then turn it over. If you do this activity on a beach, make sure you take the bandages with you afterwards and dispose of them carefully. You can leave the sand behind, though!

Time
15 minutes

Difficulty
Easy

WHAT YOU NEED

Water

Trowel

Bandages

Scissors

Bucket

Sand

Heavy books

1 Pour enough water into the sand to make the sand slightly damp. Stir the mixture thoroughly, so that there is no dry sand and no excess water.

2 Use the trowel to make a layer of sand about 5 cm (2 in) deep in the bottom of the bucket. Spread the sand evenly across the bucket's base.

3 Press down firmly on the damp sand with your hand, to compact it. Make sure the layer is flat.

As you start to build your sandcastle and press down, you force the sand grains together.

4 Carefully cut several strips of bandage that are about as long as your bucket is wide. You can always cut more if you run out of pieces.

The number of layers depends on the size of your bucket.

Bandages are made of a thin but strong mesh of woven fabric.

5 Put a few strips of bandage on top of the layer of sand. Place them so that they overlap slightly, in order to cover the sand completely.

6 Keep adding layers of damp sand 5 cm (2 in) deep, with strips of bandage between them. Press down each layer firmly.

7 Don't worry too much if your bucket is an odd shape, like this one. Just cover as much of each sand layer as possible.

8 Fill the bucket to the top with sand. Cover this final layer with strips of bandage that extend slightly over the edge of the bucket.

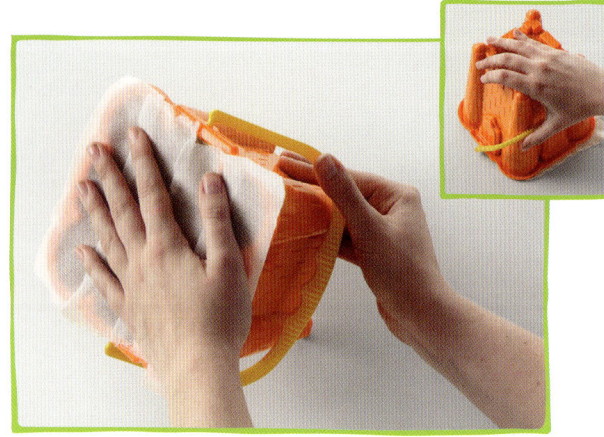

9 With one hand over the bandages to stop them falling off, carefully turn the bucket upside down and place it on the ground or a table.

Lift up your bucket slowly!

10 Gently tap the sides of the bucket, and then lift it away, just as you would if you were making an ordinary sandcastle.

12 Load more books on top of your sandcastle. See how heavy a load your sandcastle can support before it collapses!

11 Your sandcastle is ready to test! To see how sturdy it is, gently place one of the heavy books on top.

Place the books carefully on top of your sandcastle!

Engineers use the word "loading" to describe the forces a structure has to withstand.

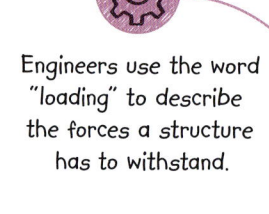

This is what happens to a normal sandcastle when you put books on it!

TAKE IT FURTHER

Your sandcastle should be able to support quite a heavy load, but can you make it even stronger? What happens if you replace the strips of bandage with paper, plastic bags, or bits of an old T-shirt? Does coarse or gritty sand make a better sandcastle than fine sand? Could you make a castle from gravel instead of sand?

HOW IT WORKS

Sand grains are made of rock and shells that have been broken down into tiny pieces by the action of moving water in the sea or in rivers. An ordinary sandcastle would collapse if you placed a load on top of it, because the sand grains can easily slip sideways over each other. But your sandcastle can support a load because the bandages increase the friction between the grains as they slip sideways. Friction is the resistance created when two or more objects are pushing past each other. Increasing the friction between the bandage strips and the grains prevents the grains from slipping sideways.

The heavy load pushes the sand grains together.

With reinforcement, the sandcastle is able to support a heavy load.

Without reinforcement, there is little friction, or resistance. The sand grains slip sideways, collapsing into a pile.

The strips of bandage increase the amount of friction in the structure, so the sand grains can't slide sideways as easily.

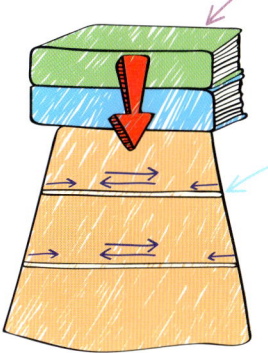

ORDINARY SANDCASTLE   REINFORCED SANDCASTLE

REAL WORLD: ENGINEERING
STABILIZING SLOPES

The combination of sand and strips of bandages used to build your sandcastle is similar to a technique used by engineers to create reinforced structures. They use grainy materials like soil or sand and pack them between strips of mesh. Where a motorway carves through the land, engineers use this technique to reinforce unstable slopes beside the road. Seawalls protecting coastal areas are also built in this way, as they are able to absorb the impact of waves and prevent coastal erosion.

LIGHT AND SOUND

Waves are associated with water, but some types of energy, such as light and sound, also travel in waves. In this chapter, you'll learn all about waves by making your own wave machine. You'll also study light with a scientific device called a spectroscope. The sounds we hear are made by waves of vibrations that disturb the air. You'll make musical sounds, by creating your own harmonica and guitar, and hear the sound of bells – with spoons!

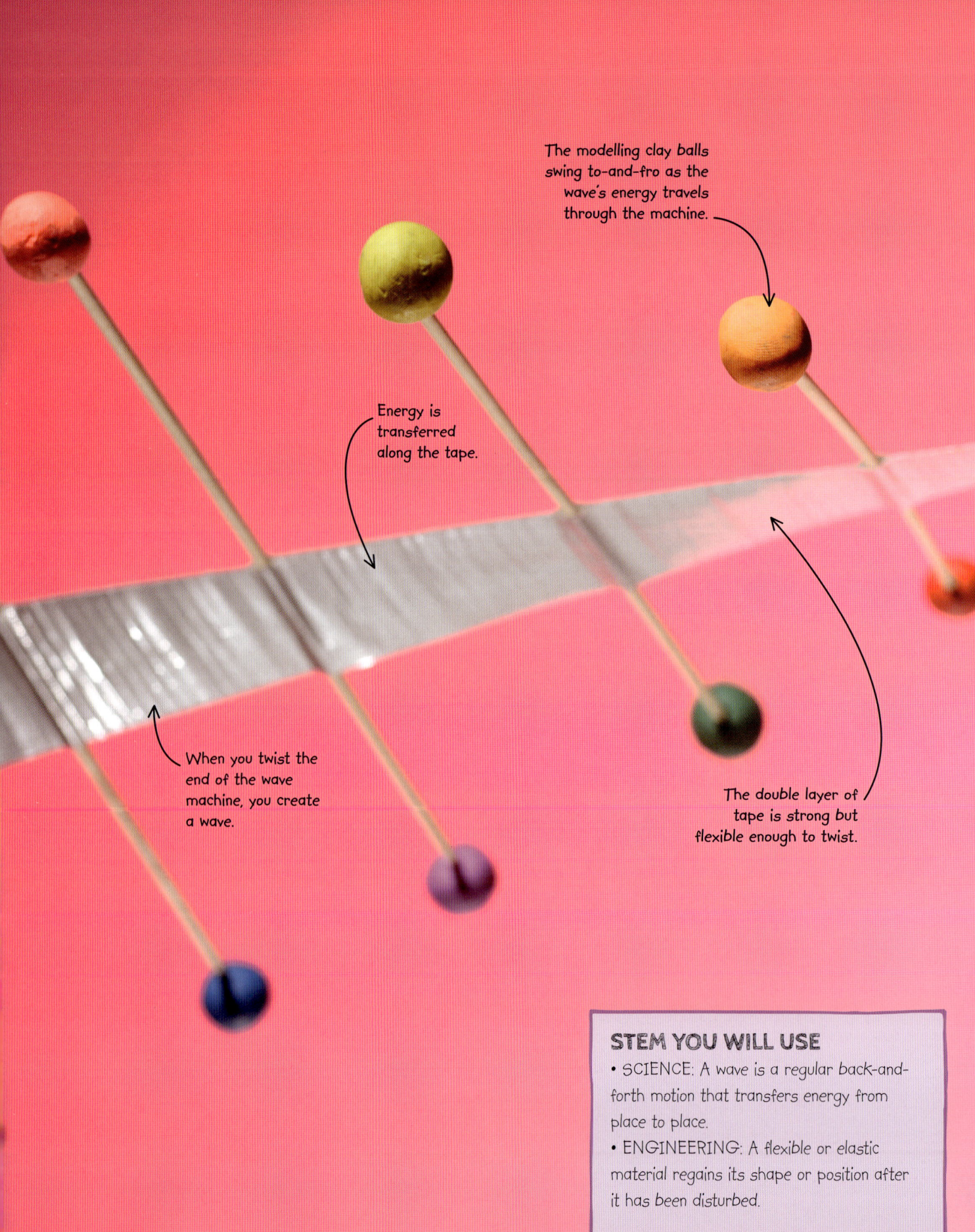

The modelling clay balls swing to-and-fro as the wave's energy travels through the machine.

Energy is transferred along the tape.

When you twist the end of the wave machine, you create a wave.

The double layer of tape is strong but flexible enough to twist.

STEM YOU WILL USE
- SCIENCE: A wave is a regular back-and-forth motion that transfers energy from place to place.
- ENGINEERING: A flexible or elastic material regains its shape or position after it has been disturbed.

The tape and the skewers twist all the way along the machine (and back again).

WAVE MACHINE

Throw a stone in a pond and it makes waves. It might look like circles of water are rushing outwards, but in fact the water is only moving up and down as energy transfers through it. Waves transfer energy from place to place, which makes them very useful. We can use them to send and receive information, to heat up food, and to surf! This twisting wave machine lets you see waves in action.

HOW TO MAKE A
WAVE MACHINE

You'll make your wave machine from duct tape and wooden skewers. The tape is very sticky and can easily become stuck to itself, so take your time. The skewers have a sharp point at one end, so be careful. You'll need to work somewhere with plenty of space because your wave machine will be about 3 m (9 ft) long!

Time
30 minutes

Difficulty
Easy

WHAT YOU NEED

Scissors

Measuring tape

Duct tape

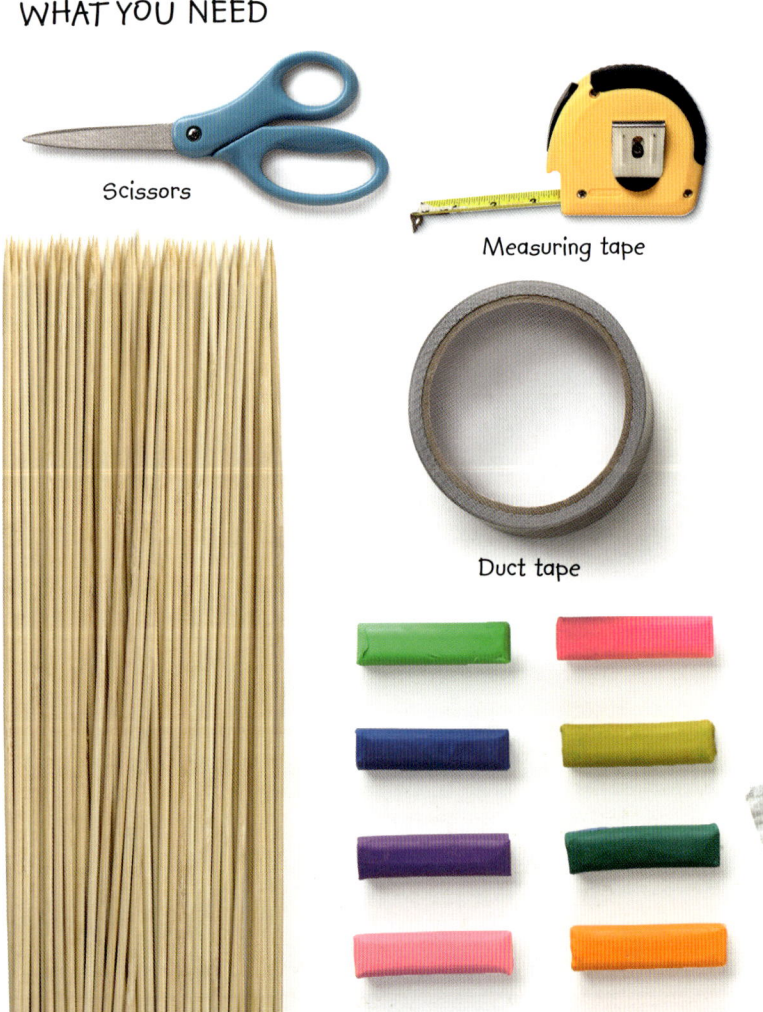

Wooden skewers

Modelling clay

First you'll make the handles. Carefully cut a length of tape about twice as long as a skewer. Lay it down on the table sticky side up.

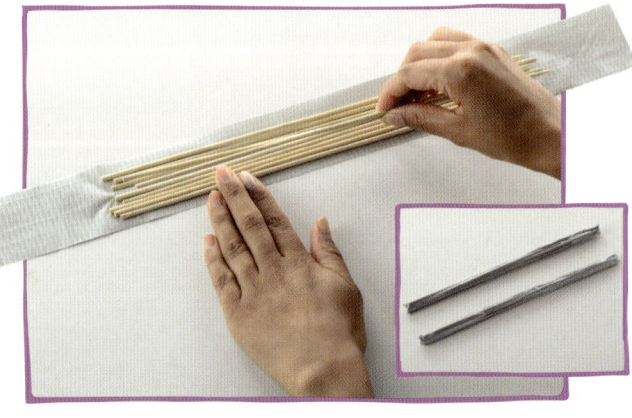

2 Put 10 skewers onto the tape's centre. Roll the tape tightly round the skewers and fold the ends over, then repeat to make another handle.

3 Pull out some of the tape, with the sticky side facing upwards. Place one handle onto the tape and roll it forwards a short way to secure it.

4 Roll out about 1 metre (3 ft) of tape. Working from the end nearest the handle, place a skewer across the tape every 5 cm (2 in) or so.

5 Roll out another metre (3 ft) of tape and keep adding a skewer every 5 cm (2 in). Repeat this step so your wave machine is about 3 m (9 ft) long.

Leave about 10 cm (4 in) between the last skewer and the handle.

6 When the wave machine is 3 m (9 ft) long, roll out another 20 cm (8 in) of tape and cut. Place the other handle on top of the tape and roll it for 10 cm (4 in) leaving a 10 cm (4 in) gap without skewers.

7 Starting at the second handle, place a new strip of tape sticky side down onto the skewers, on top of and all the way along the upturned tape.

8 Now roll lots of small round balls of modelling clay of a similar size – two for each wooden skewer you've placed on the tape.

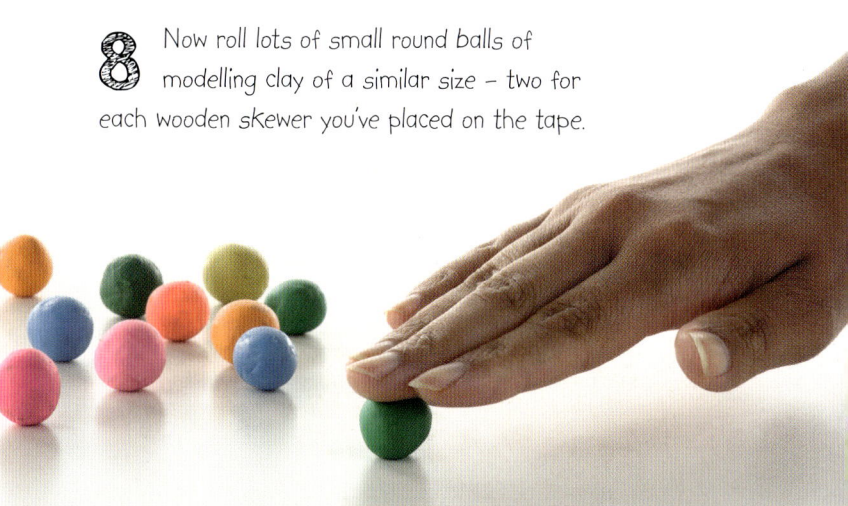

9 Push one ball of modelling clay onto each end of every skewer. Be careful not to stab your fingers on the sharp ends of the skewers.

10 Your wave machine is ready to use! Ask a friend to hold one handle completely still or secure the handle to a piece of furniture. Hold the other handle, stretching it out gently. Now twist it by alternately lifting your left and right hand.

TEST AND TWEAK

Can you make different waves by changing the design of your wave machine? You could try moving the wooden skewers closer together or further apart and making the modelling clay balls smaller or larger – or try missing some of them off altogether! You could also work out the speed of your waves in metres or feet per second by measuring the tape, timing how long a wave takes to travel along it, and dividing the distance by the time.

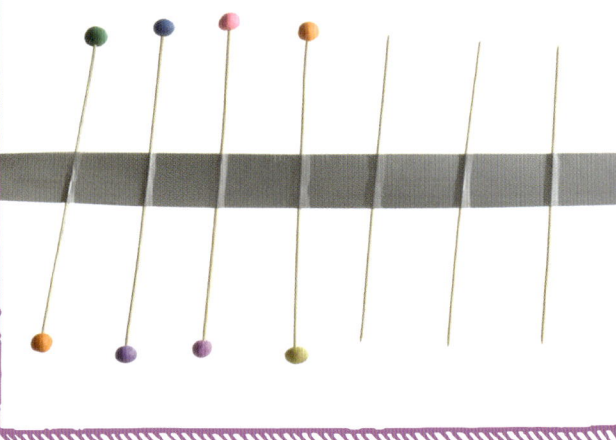

The energy you use twisting the handle is transferred along your wave machine.

HOW IT WORKS

Waves are all around us and take many different forms, from ripples in water to sound waves, light, and radio waves. Your wave machine shows how waves transfer energy from place to place. When you twist the handle, the energy you use transfers along the wave machine. Only the energy moves forwards – the modelling clay balls merely wobble back and forth as they pass on the energy. All types of wave have a speed, a wavelength, and a frequency. The speed is how fast the wave travels. The wavelength is the distance between two "peaks" or "troughs". The frequency is how many waves pass a particular point every second.

LOW FREQUENCY

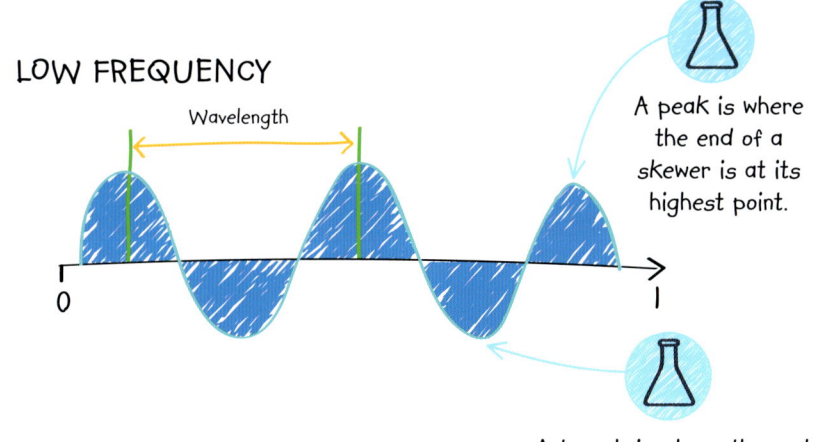

A peak is where the end of a skewer is at its highest point.

A trough is where the end of a skewer is at its lowest.

HIGH FREQUENCY

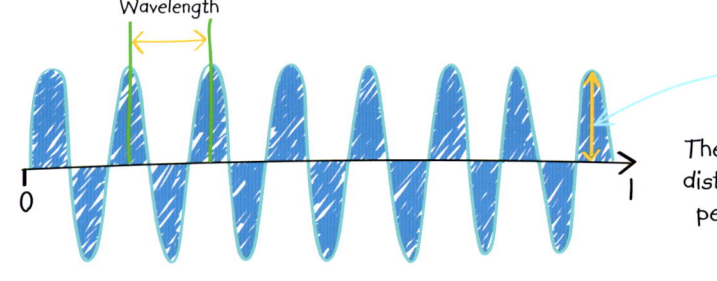

The amplitude is the distance between the peak and midpoint of a wave.

REAL WORLD: TECHNOLOGY
FIBRE OPTIC CABLES

Much of the information on the internet is shared around the world at high speed by waves of infrared radiation. These light waves pass through thin glass threads known as optical fibres. The light waves reflect and bounce off the insides of the glass threads, travelling along without losing much of their energy, and transmitting data over huge distances.

Light from a torch appears white, but it is a mix of many different colours. It enters through a slit at the top of the spectroscope.

Daylight is a good source of white light if you don't have a torch available.

When white light hits the shiny side of the CD, it bounces off and separates into different colours.

A viewing window lets you see and study the spectrum, or range, of colours present in the torchlight.

SPECTROSCOPE

It might look white, but light is actually a mix of different colours. Scientists use a device known as a spectroscope to study the range of colours (the spectrum) in different kinds of light. In this activity, you can make your own spectroscope.

STEM YOU WILL USE
• SCIENCE: Colour spectrums vary depending on the source.
• TECHNOLOGY: Spectroscopes analyse light by separating it into the colours it's made of.

HOW TO BUILD A
SPECTROSCOPE

In order to clearly see the spectrum of colours that make up white light, you'll need a shiny CD for the light to bounce off. A slit at the top of a dark tube lets a small amount of light into the tube and onto the CD. You'll need to use a protractor to measure the angle at which you place the CD. You'll also need black electrical tape to block out unwanted light.

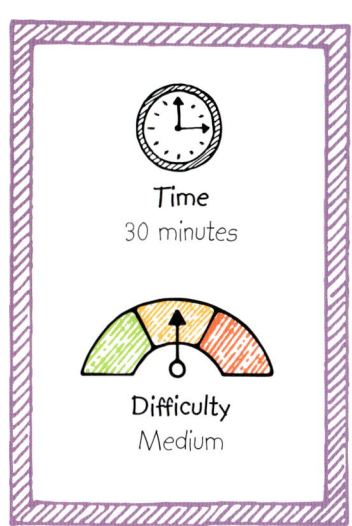

Time
30 minutes

Difficulty
Medium

WHAT YOU NEED

Protractor

Black electrical tape

Paint

Scissors

Torch

Paintbrush

Pencil

Black card

Old CD or DVD

Cardboard tube

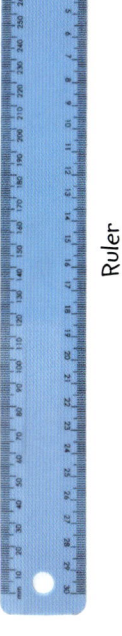

Ruler

3 cm
(1¼ in)

1 Using the pencil, make a mark 3 cm (1¼ in) from one end of the cardboard tube.

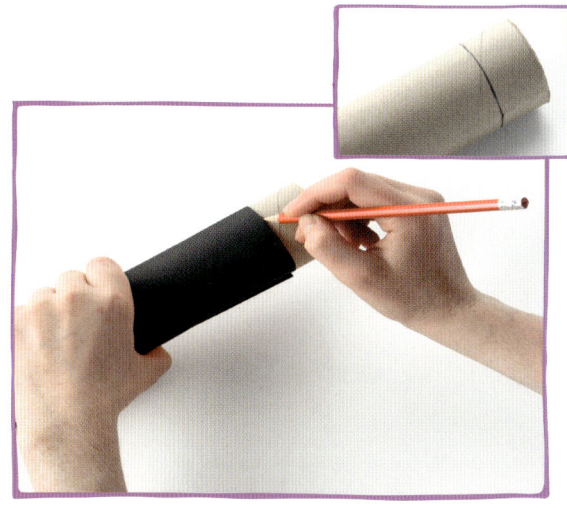

2 Wrap the black card around the cardboard tube at the mark. Use it as a guide to draw a line around the tube.

Check the straight edge of your protractor is still lined up with the pencil line you drew.

3 Hold the protractor on the tube so the protractor's zero line runs along the pencil line. Draw a short line angled at 30°.

4 Move the protractor and draw another line, angled at 30° in the other direction, so the two slanted lines almost meet.

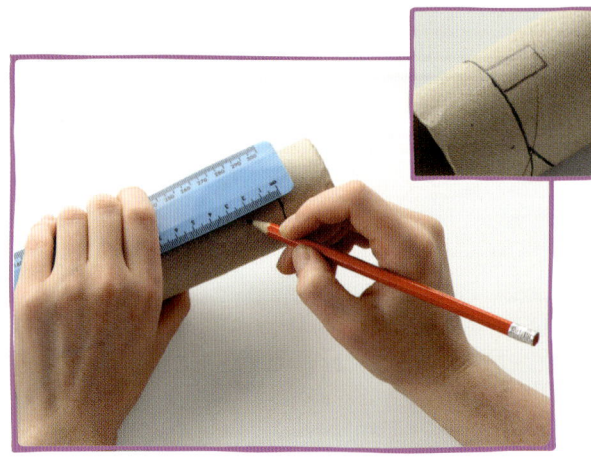

5 Using the ruler, extend both slanted lines so they meet the line that goes around the cardboard tube, forming a triangle.

6 On the opposite side of the tube from the triangle, draw a rectangle 2 cm (¾ in) high and 1 cm (½ in) wide above the pencil line.

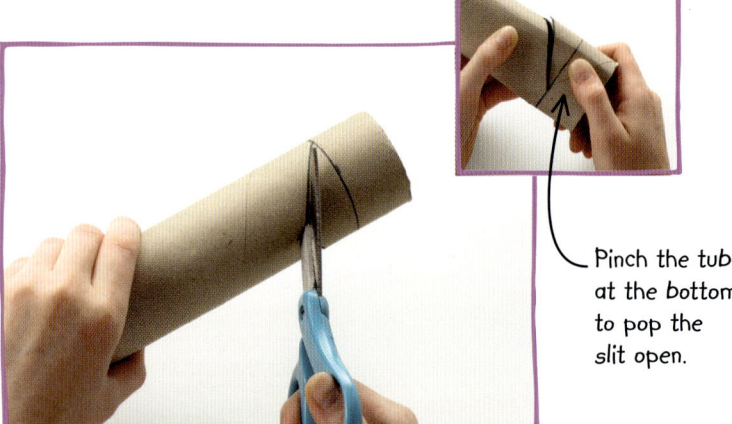

Pinch the tube at the bottom to pop the slit open.

7 Cut along the two slanted lines so that you end up with an angled slot. This is where you'll slide in your old CD.

8 Now carefully cut out the small rectangle you drew, to make a viewing window for your spectroscope. Ask an adult to help if you get stuck.

9 Paint the cardboard tube in any colour or design you like, then leave the paint to dry.

11 Secure the CD in place inside the slot using black electrical tape.

10 Push the CD into the angled slot, with the shiny, bottom surface facing upwards.

It's crucial to get the angle of the CD just right so you can see the spectrum clearly.

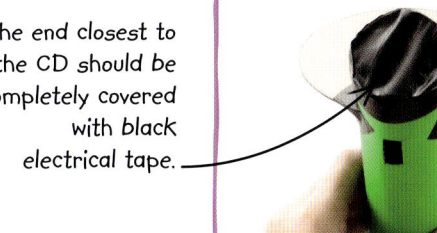

12 Use strips of electrical tape to close off the end of the cardboard tube closest to the CD. Make sure no light can get into the tube.

The end closest to the CD should be completely covered with black electrical tape.

13 Draw around the open end of the cardboard tube onto the black piece of card, using a pencil. Carefully cut out the circle.

14 The circle of card will cover the open end of the tube, but it needs a slit to let in light. To make the slit, first fold the circle in half.

15 Carefully cut two lines close together at right angles from the middle of the fold. Then snip off the thin piece between the lines.

16 Unfold the circle and tape it over the open end of the tube. The slit should run from side to side, not front to back, so that it aligns with the slot holding the CD.

Light from the torch enters the slit in the top of the spectroscope.

Most modern torches use LEDs (light emitting diodes) as their light source, which is made up of fewer colours than those found in sunlight.

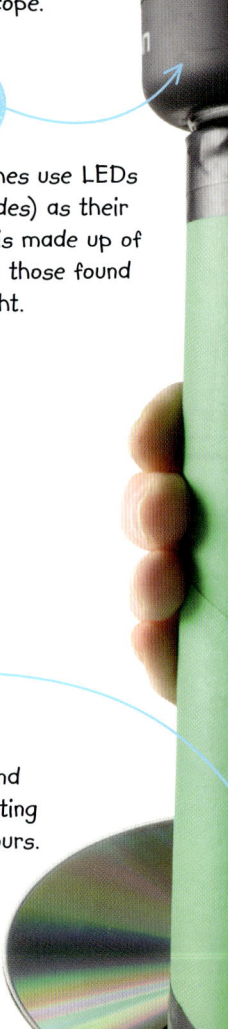

The light hits the shiny CD and bounces off, splitting into different colours.

Carefully tape around the circle of black card to hold it in place.

17 Your spectroscope is now ready to use! Shine a torch into the top and look through the viewing window to see the spectrum. You could also try other light sources, such as daylight through a window, but don't point your spectroscope directly at the Sun.

HOW IT WORKS

White light is a mixture of all the colours of the rainbow. When it hits a reflective object, all these colours bounce off, or reflect. Light hitting the shiny underside of a CD reflects in a different way. All the colours reflect, but each one bounces off in a different direction. The different colours spread out to form a spectrum.

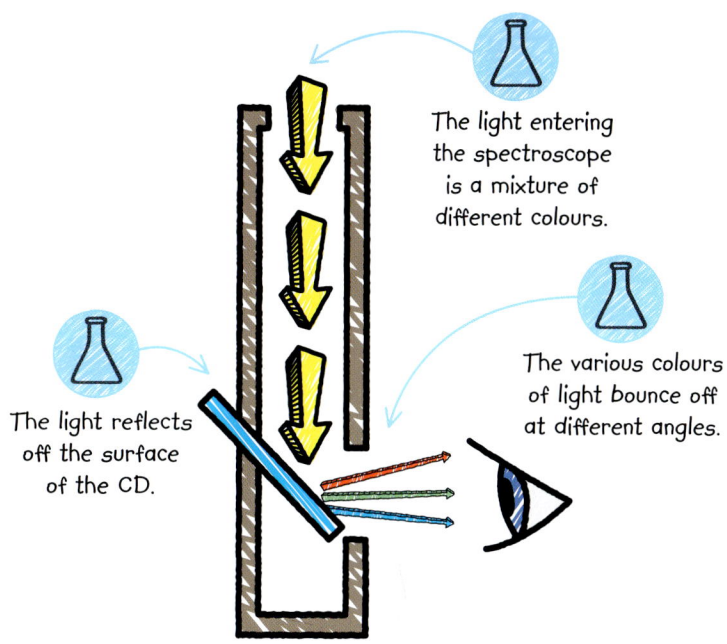

The light entering the spectroscope is a mixture of different colours.

The various colours of light bounce off at different angles.

The light reflects off the surface of the CD.

COMPARING DIFFERENT LIGHT SOURCES

If you compare different light sources, such as daylight or the screen of a mobile phone, you'll find that each one produces a distinctive spectrum. Daylight produces a continuous spectrum, with every colour of the rainbow and no gaps. In contrast, an artificial light source typically produces only certain colours, and so its spectrum has coloured lines with black gaps between them.

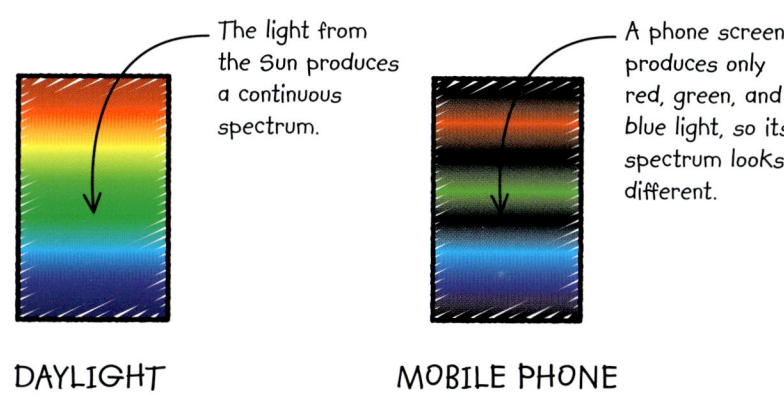

The light from the Sun produces a continuous spectrum.

A phone screen produces only red, green, and blue light, so its spectrum looks different.

DAYLIGHT

MOBILE PHONE

REAL WORLD: SCIENCE
THE LIGHT OF STARS

Each of the chemical elements of which matter is made produces light with a different spectrum when it burns. Chemists in laboratories can identify which elements are present in different substances by using spectroscopes to study the light they give off when burned. Astronomers also use spectroscopes to study light from stars; from lines in the spectrum, they can tell which elements are present.

SINGING SPOONS

In this activity, you'll use metal spoons to create amazing noises that sound like bells or gongs chiming – but you'll only hear these incredible sounds if you put your fingers in your ears! When the spoons swing and bang together, the metal flexes a tiny amount, and then flexes back again repeatedly. These movements, known as vibrations, are too fast and small to see, but they cause the string attached to the spoon to vibrate as well. The vibrations pass along the string and ultimately into your ears!

STEM YOU WILL USE
• SCIENCE: Sound is created when objects vibrate. Sound travels through air, but also through solids, such as pieces of string.

You'll need to place your fingers gently into your ears to hear the sounds!

Vibrations pass along the string and through your fingers.

Hanging the spoons from a string means they are able to vibrate freely.

The spoons clink as they knock together, but when you put your fingers in your ears, they sound different.

HOW TO MAKE
SINGING SPOONS

To hear the amazing sounds that spoons can produce, all you need to do is secure three metal spoons to string, wrap the string around your fingers, put your fingers in your ears, and knock the spoons together. This activity is super simple and quick to do, but you'll be surprised by the results!

Time
10 minutes

Difficulty
Easy

WHAT YOU NEED

String

Sticky tape

Scissors

Three metal spoons

This string will carry the vibrations made by the spoons as they knock together.

1 Cut a piece of string that is about twice the length of your arm. Lay it flat on a table.

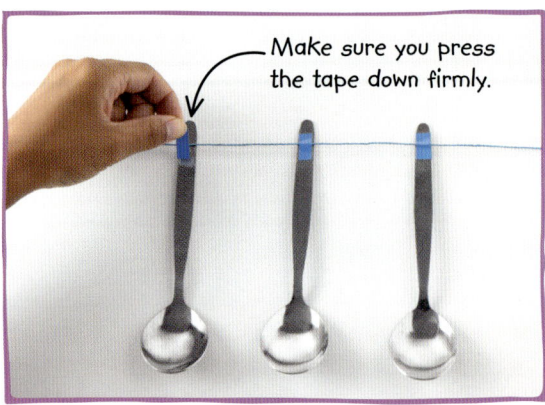

Make sure you press the tape down firmly.

2 At the string's middle point, place the ends of the spoons a few centimetres apart. Secure each spoon to the string with a piece of tape.

Metal spoons are sonorous, which means they make a ringing sound, like a bell.

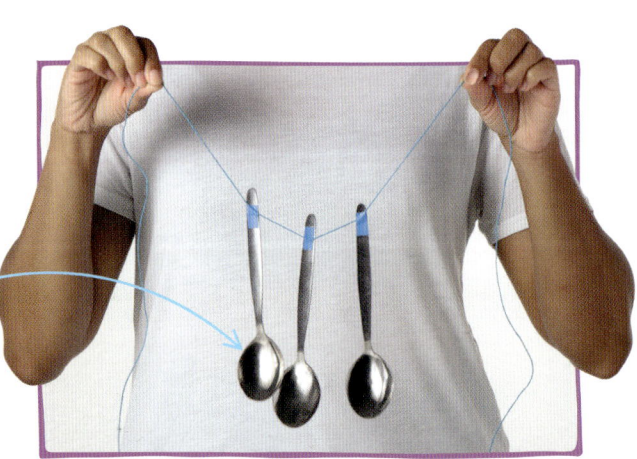

3 Dangle the spoons in front of you and wrap the string around one finger of each hand. Shake the spoons so they knock each other – they'll make a sharp, tinny sound.

Putting your fingers in your ears blocks out other noises, so the sound of the spoons seems even louder.

4 Put the fingers with string around them in your ears. Knock the spoons together – the spoons will sound louder and richer, like bells chiming.

TAKE IT FURTHER

If you switch the metal spoons for other metal objects, such as keys, or nuts and bolts, how does the sound change? Does the experiment still work if you use wooden or plastic spoons instead?

HOW IT WORKS

When metal spoons knock together, they vibrate (move rapidly to and fro). This makes sound because the vibrations make air molecules vibrate too, creating invisible waves that travel through the air to your ears. Sound waves spread out as they travel through air, so the sound you hear is quiet. But the vibrations also pass through solid materials (the string, your fingers, and your skull, which houses your inner ears). Since these vibrations do not spread out, the sound is louder when it reaches your ears. Sound waves travel more effectively through solids than air because the molecules are more tightly packed together. As a result, you hear a richer, more complex pattern of sound waves.

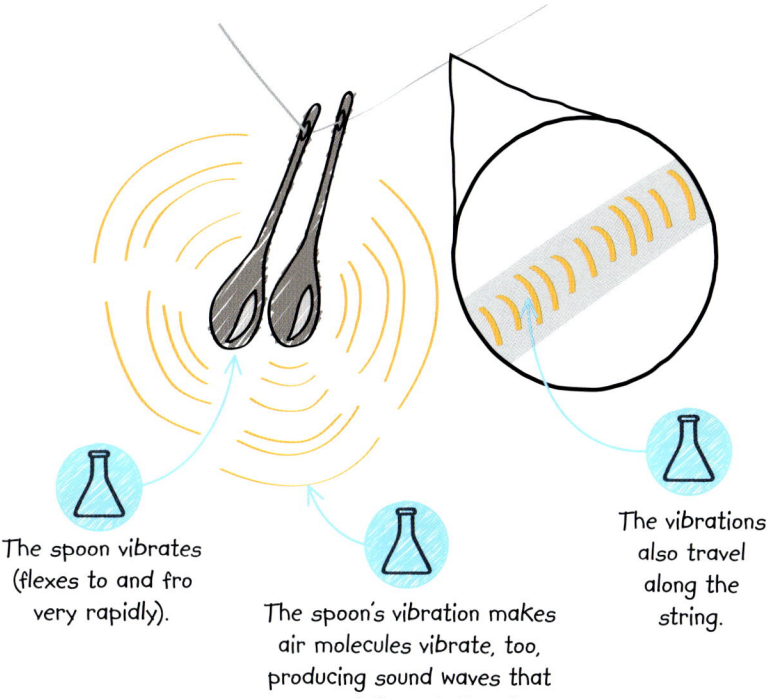

The spoon vibrates (flexes to and fro very rapidly).

The spoon's vibration makes air molecules vibrate, too, producing sound waves that spread through the air.

The vibrations also travel along the string.

REAL WORLD: TECHNOLOGY
STETHOSCOPE

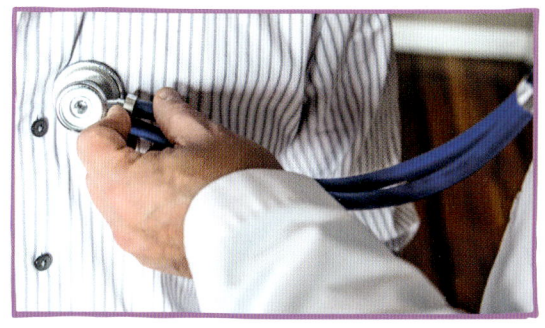

You can't normally hear your heartbeat as it's too quiet. However, a doctor can hear it with a device called a stethoscope. At one end is a cup that the doctor presses on your chest to collect the faint sound of the heart beating. A hollow tube channels this sound to the doctor's ears, preventing the sound waves from spreading out in all directions and becoming too faint to hear.

HARMONICA

Here's a fun and simple way to make some musical notes, and to learn a bit about the science of sound. Just like a real harmonica, this one has a part that vibrates when you direct air past it with your breath. In this homemade version, it's a piece of paper held between two pieces of cocktail stick sandwiched between two lollipop sticks. Go on – see what weird and wonderful sounds you can make!

The pitch of the note your harmonica produces – how high or low it is – depends on how fast the paper vibrates. The faster it vibrates, the higher the pitch.

STEM YOU WILL USE

• SCIENCE: Sound is produced by vibrating objects. Musical instruments produce notes that vibrate at either a high or low pitch. By stretching a vibrating rubber band you can produce a higher-pitched sound.

Making harmonicas is a good way to recycle your lollipop sticks, but make sure they are dry first.

As you play the harmonica, you'll be able to feel the tickle of the vibrations that create the sound.

HOW TO MAKE A
HARMONICA

This harmonica is made with lollipop sticks. As you are going to be touching these with your mouth, make sure they are clean. The only other things you need are rubber bands, toothpicks, and a strip of paper. You'll be making music in just a few minutes!

Time
15 minutes

Difficulty
Easy

WHAT YOU NEED

Two rubber bands

Two toothpicks

Pencil

Scissors

Coloured paper

Two lollipop sticks

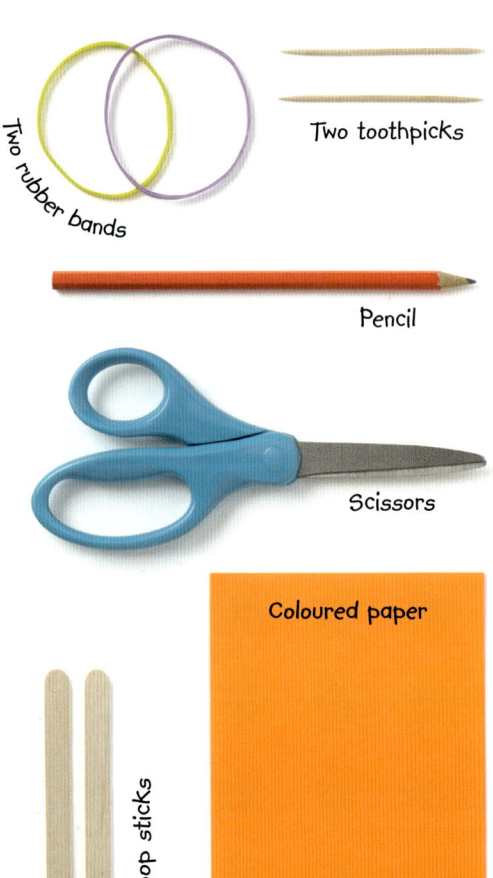

1 On the paper, draw around a lollipop stick with the pencil, then carefully cut around the shape with the scissors.

In your harmonica, the paper vibrates, disturbing the air around it to produce sound.

2 Place the piece of paper you cut out on top of one of the lollipop sticks, then place the other lollipop stick on top.

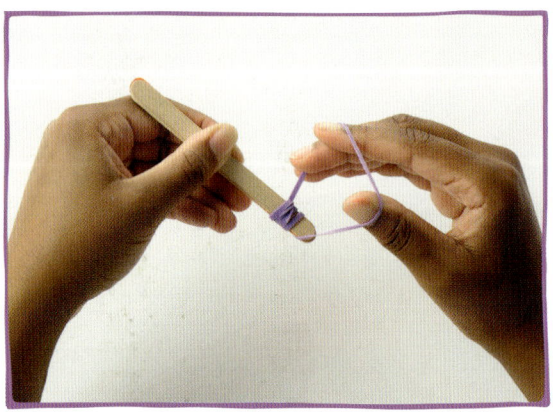

3 Wrap a rubber band several times around one end of the two lollipop sticks, so that it holds them together.

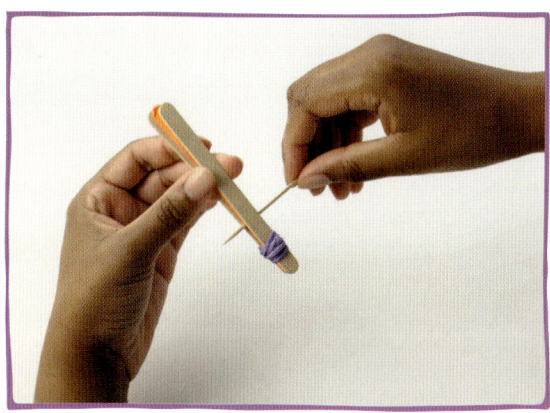

4 Wedge a toothpick between the lollipop sticks, and slide it as far towards the end with the rubber band as you can.

You should end up with a toothpick wedged in at both ends of the harmonica.

5 Wrap another rubber band around the other end of the lollipop sticks, then wedge another toothpick in at that end.

6 Using scissors, carefully trim the toothpicks and discard the extra pieces. Make sure the paper is flat, not crumpled, then hold the harmonica firmly between your lips and blow. Try sucking, too.

What happens when you press the lollipop sticks together with your teeth?

HOW IT WORKS

The pieces of toothpick hold the paper firmly at each end. When you blow or suck, air rushing past the paper makes it vibrate, and the vibrations create disturbances in the air that travel outwards in all directions as sound waves. If you blow harder or pinch the sticks as you blow, the paper vibrates faster. This creates a higher-pitched sound.

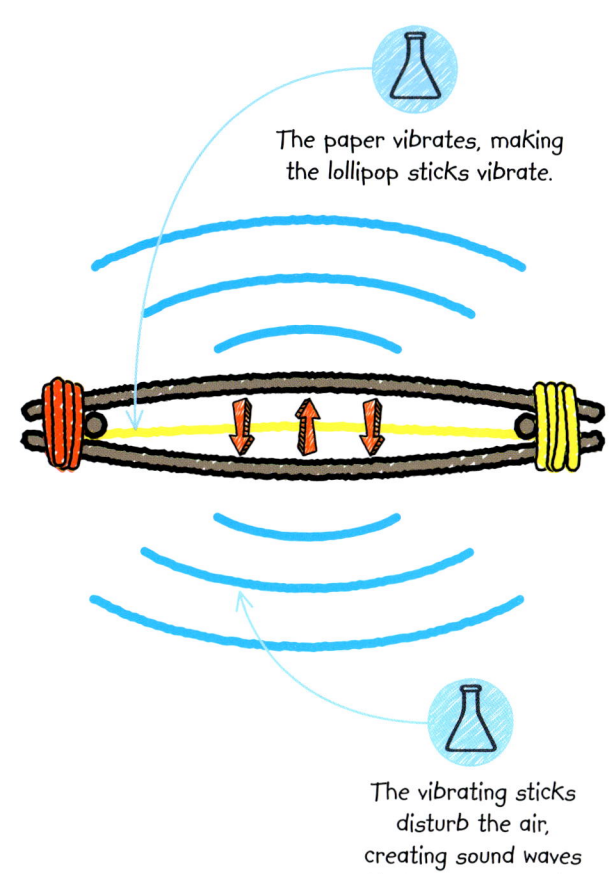

The paper vibrates, making the lollipop sticks vibrate.

The vibrating sticks disturb the air, creating sound waves that move outwards.

REAL WORLD: SCIENCE
VIBRATING REEDS

Real harmonicas work in a similar way to your lollipop harmonica. Instead of paper, they have metal sheets called reeds that vibrate when the player blows or sucks through a set of holes. There is at least one reed behind each hole, and each reed is tuned to a different note.

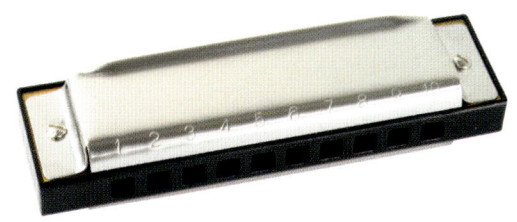

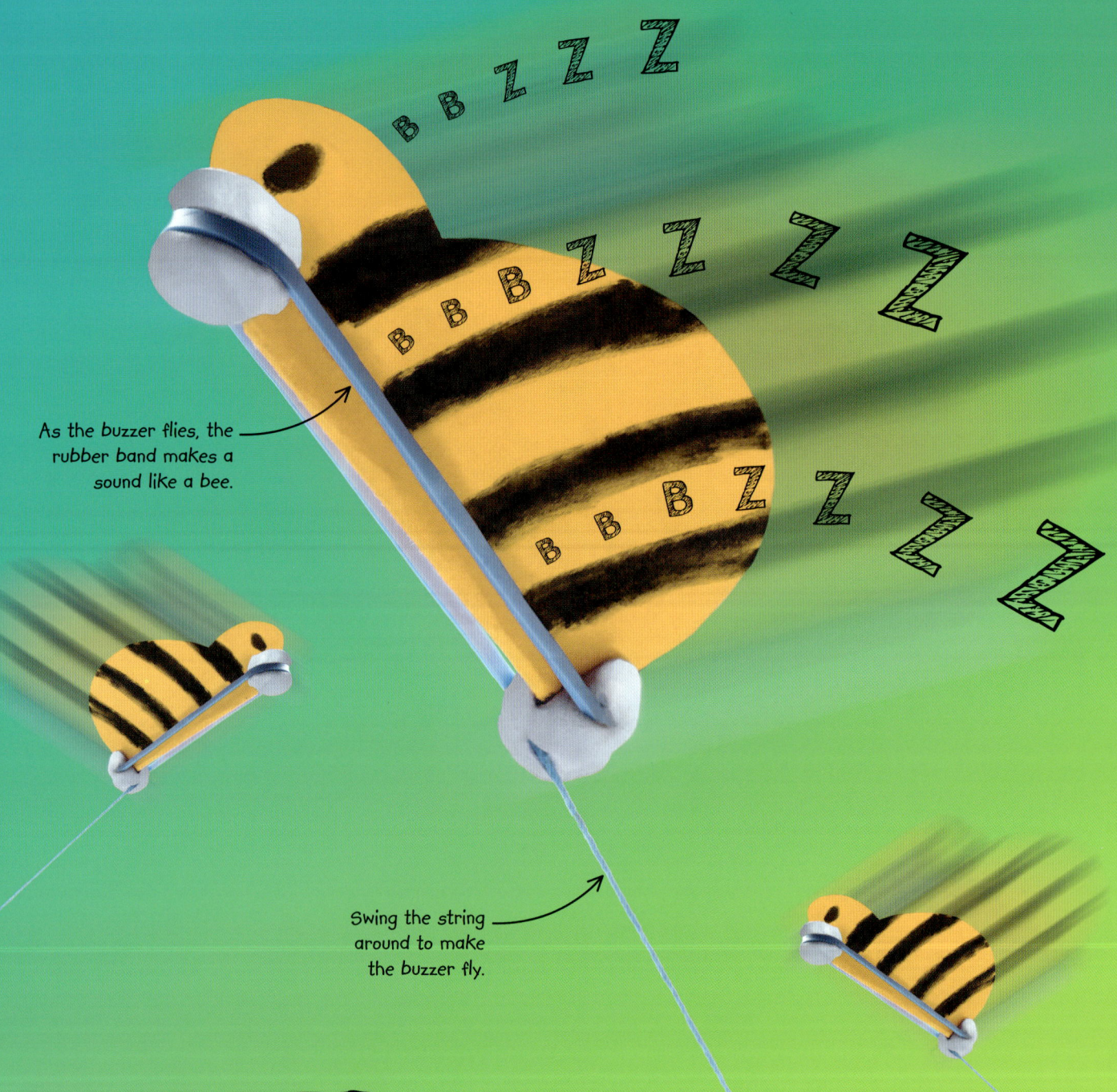

As the buzzer flies, the rubber band makes a sound like a bee.

Swing the string around to make the buzzer fly.

BUZZER

When you whirl this buzzer around, it'll make a sound like a bumblebee. A bee makes a buzz by flapping its wings more than 200 times a second when it flies. Instead of wings or muscles, your buzzer will use a simple rubber band to mimic a flying bee. The rubber band twists back and forth rapidly as you whirl it through the air, and this fluttering movement creates sound waves.

HOW TO MAKE A
BUZZER

The buzzer is made from a lollipop stick, a rubber band, some card and string, and some adhesive putty. It's quick and easy to make – but you may have to adjust certain things to make your buzzer work well. In particular, you might have to try a few different sizes of rubber band.

Time
15 minutes

Warning
Take care not
to hit yourself
or anyone else.

Difficulty
Easy

STEM YOU WILL USE

• SCIENCE: The force needed to keep something moving in a circle is known as centripetal force. Also, air flow over a flexible object can cause it to vibrate – a phenomenon known as aeroelastic flutter.

WHAT YOU NEED

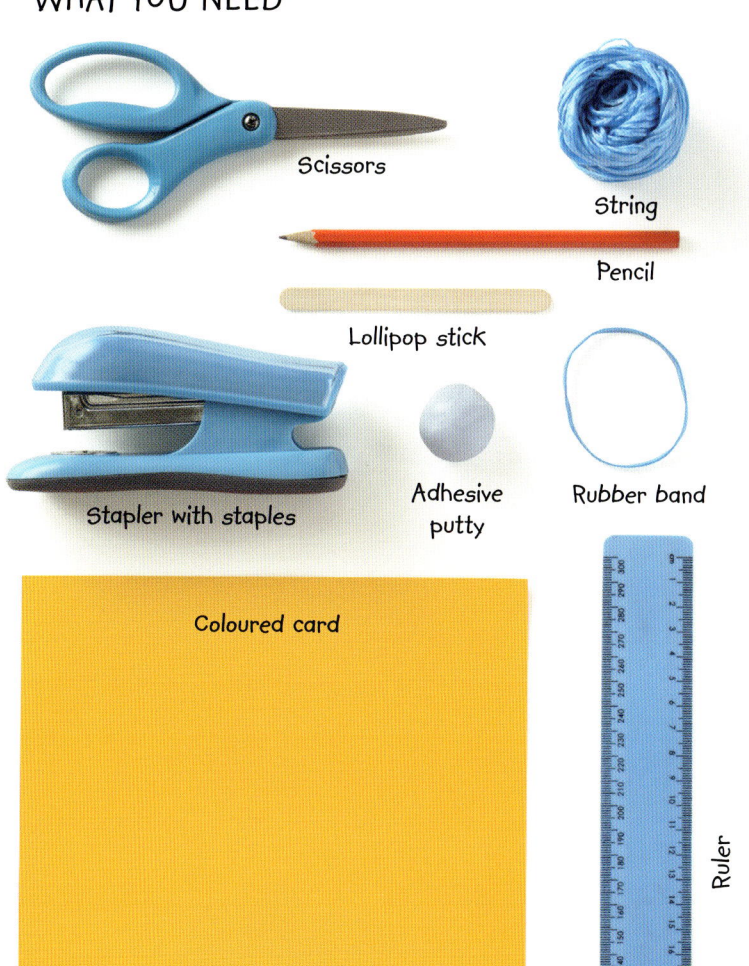

Scissors

String

Pencil

Lollipop stick

Stapler with staples

Adhesive putty

Rubber band

Coloured card

Ruler

1 Fold the card in half. Make a tight crease by pressing down firmly along the fold.

Folded edge

2 Lay the lollipop stick next to the folded edge. Make two pencil marks on the fold, each about 1 cm (½ in) from the end of the stick.

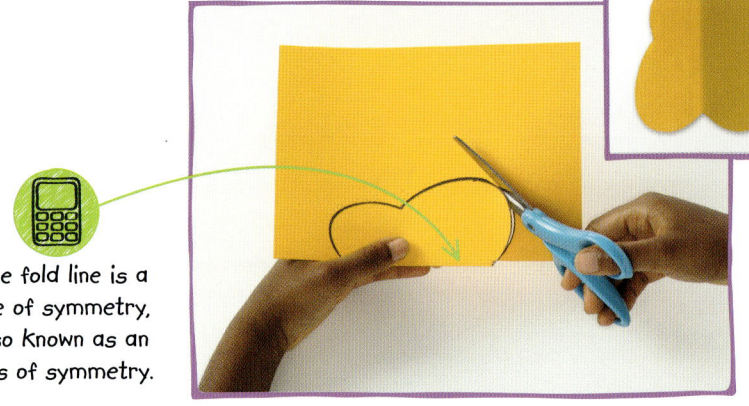

The fold line is a line of symmetry, also known as an axis of symmetry.

3 Draw a curve *between* the two pencil marks, like the shape shown here. This will form the outline of the bee's body.

4 Cut along the pencil line. If you open up the folded piece of card, you'll have a bee shape that's symmetrical.

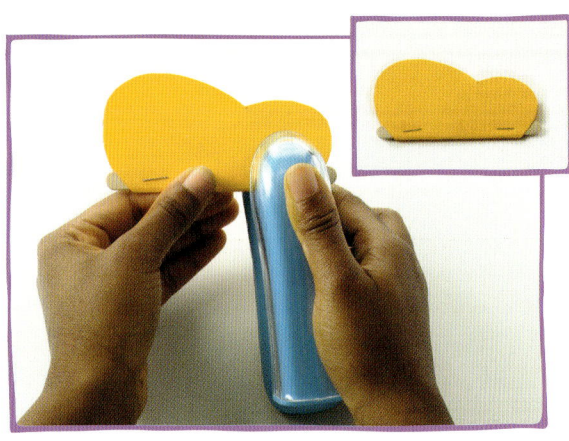

5 Place the lollipop stick inside the fold and staple through the card and the stick twice to hold the stick firmly in place.

6 Cut a length of string at least 50 cm (20 in) long and tie one end securely to one end of the lollipop stick.

7 Secure a lump of adhesive putty at each end of the lollipop stick. Make sure you press it down firmly.

Make sure the adhesive putty is firmly stuck to the lollipop stick.

If you want, draw black stripes and eyes on the card to make the buzzer look bee-like.

The elastic band should be taut enough that it doesn't fall off.

8 Stretch the rubber band over the adhesive putty lumps. Check the band isn't twisted and its sides are parallel to but not touching the lollipop stick.

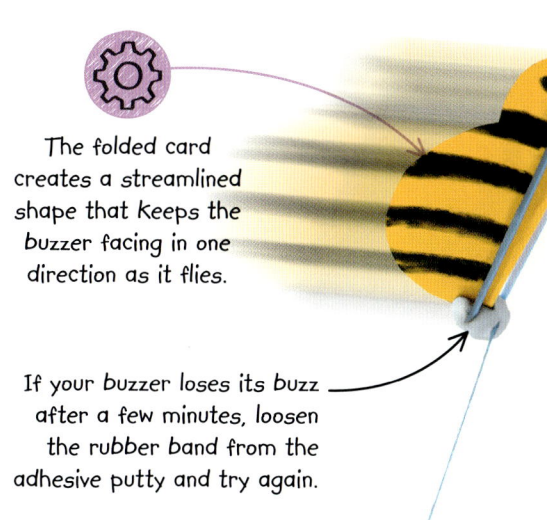

The folded card creates a streamlined shape that keeps the buzzer facing in one direction as it flies.

If your buzzer loses its buzz after a few minutes, loosen the rubber band from the adhesive putty and try again.

9 Open out the wings slightly. Find an open space where your buzzer won't hit any objects or people and whirl it around in large circular movements as fast as you can. If the buzz is too quiet, experiment with thick and thin rubber bands until you find the one that makes the loudest buzz.

As you whirl the buzzer, centripetal force pulls on the string and makes the buzzer follow a circular path.

HOW IT WORKS

The buzzing sound is caused by something called aeroelastic flutter. This happens when a flexible object is in fast-moving air, and the air makes it flex back and forth quickly. The rubber band flexes about 200 times a second, which is about the same frequency as a bee's wings and so makes a similar sound. You can create an even louder sound from flutter by sandwiching a blade of grass between the sides of your thumbs and blowing through the gap between them.

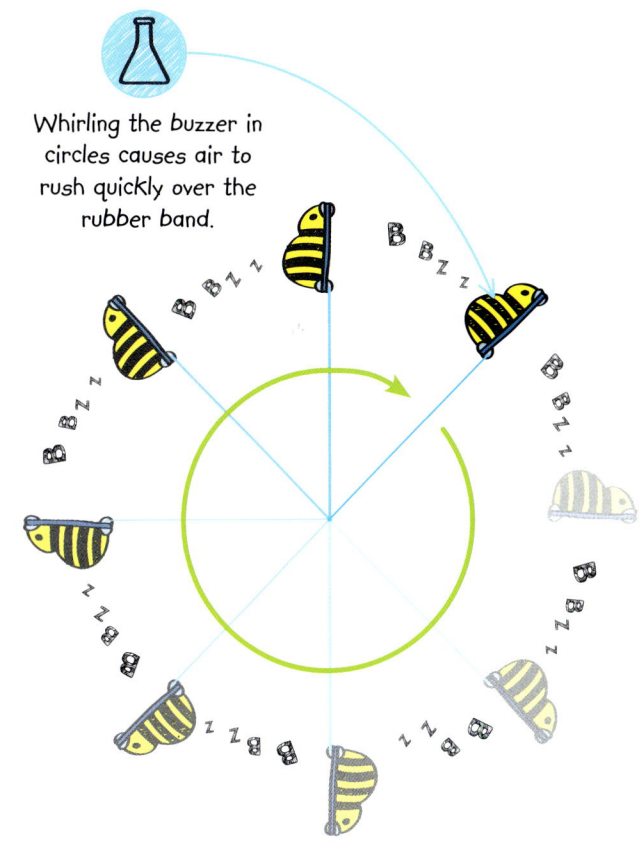

Whirling the buzzer in circles causes air to rush quickly over the rubber band.

REAL WORLD: SCIENCE
TACOMA NARROWS BRIDGE

In 1940, the world's third-largest suspension bridge (the Tacoma Narrows Bridge in the USA) was hit by strong winds and began to flutter. It twisted back and forth with such violence that it tore itself apart and collapsed. Today, engineers go to great lengths to prevent aeroelastic flutter in structures affected by fast-moving air, such as aircraft and bridges.

The strings are held in tension.

You can adjust how tightly the strings are held by turning the hooks at the head.

STEM YOU WILL USE

• SCIENCE: A vibrating string produces a note of a particular pitch. The note's pitch depends on the string's tension, thickness and length.

• TECHNOLOGY: A soundboard amplifies the sound the strings produce.

• ENGINEERING: Layering a material, such as cardboard, can give it stiffness and strength.

GUITAR

Making music can be a great way to explore the science of sound – and this guitar can help you do both. With fishing line for strings and an ice cream tub for a body, it's easy to make. And if you set it up right, it will make a surprisingly tuneful sound. In fact, you'll find this project really hits the right notes!

The vibrations of the strings are passed on to the body of the guitar.

The vibrating body of the guitar disturbs the air, sending out sound waves into the air around it.

HOW TO MAKE A
GUITAR

The two most important features of your guitar are the strings and the body. In this project, the strings are made of fishing line. The body of your guitar is made of a plastic ice cream tub. The neck is made of corrugated cardboard.

Time
1 hour and
30 minutes

Difficulty
Hard

WHAT YOU NEED

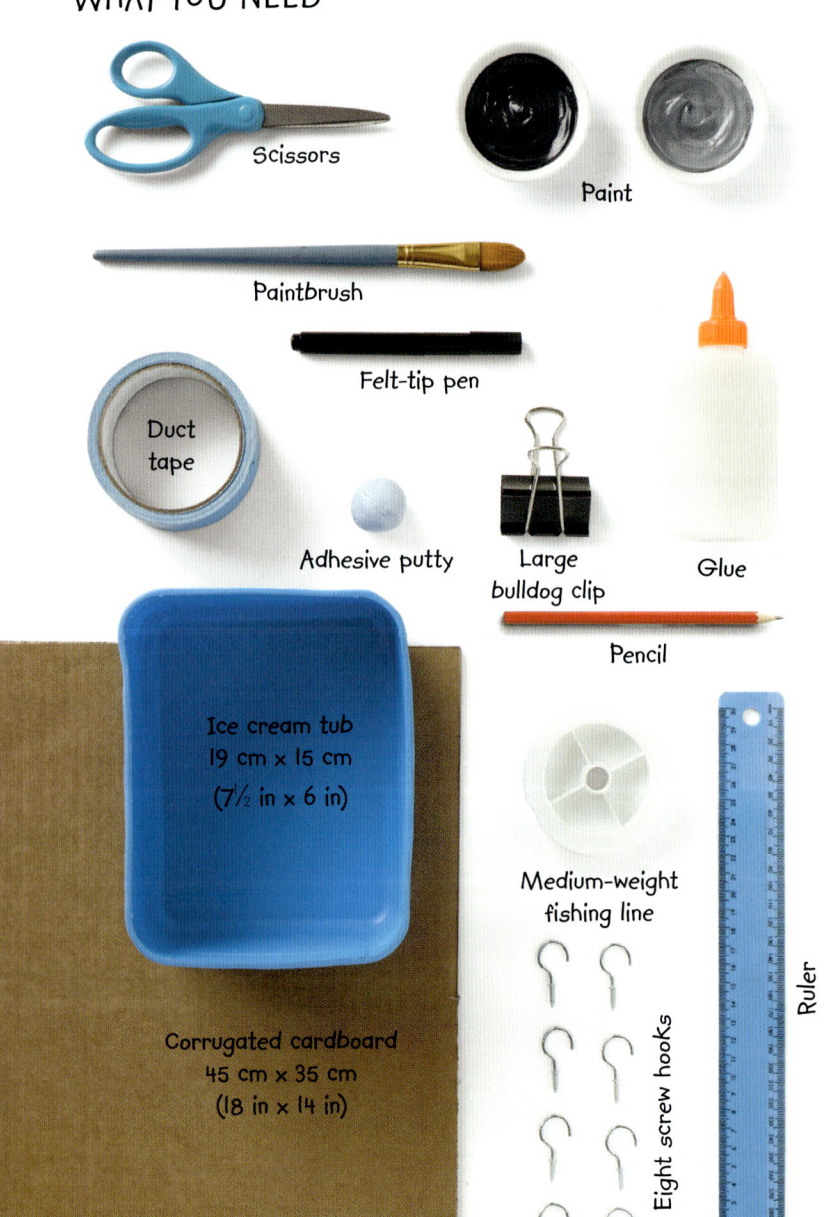

Scissors

Paint

Paintbrush

Felt-tip pen

Duct tape

Adhesive putty

Large bulldog clip

Glue

Pencil

Ice cream tub
19 cm x 15 cm
(7½ in x 6 in)

Medium-weight
fishing line

Eight screw hooks

Ruler

Corrugated cardboard
45 cm x 35 cm
(18 in x 14 in)

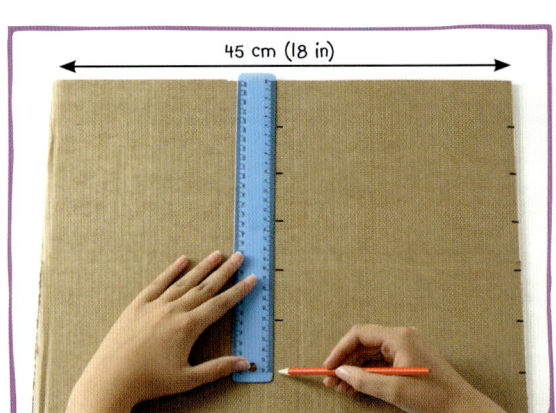

45 cm (18 in)

1 Make pencil marks every 5 cm (2 in) down the shorter side of the cardboard. Repeat the marks in the middle.

2 Using the ruler, draw straight lines that join the marks you made and extend them across the whole width of the cardboard.

3 Cut along the lines, so that you end up with seven long rectangles of cardboard, each 45 cm (18 in) long and 5 cm (2 in) wide.

4 On one of the long rectangles, make a pencil mark 22½ cm (9 in) from one end – halfway along its length.

5 Divide the long rectangle into two equal pieces, by cutting where you made the pencil mark.

Stacking and gluing the pieces together like this increases the strength of the neck.

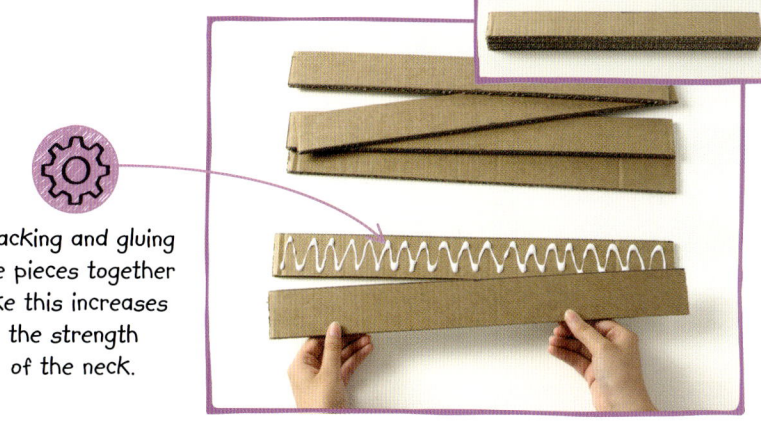

6 Stick the other six long rectangles together by applying glue between them and putting them together into a stack.

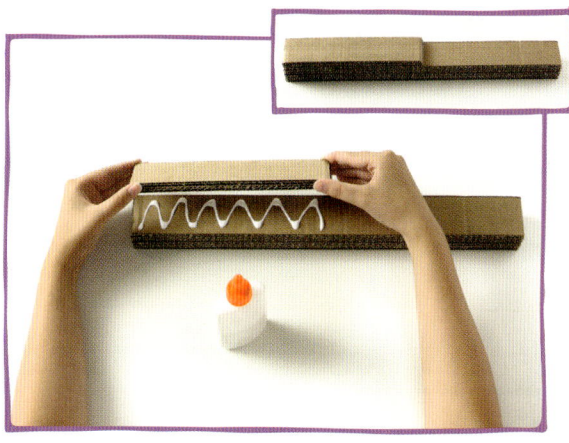

7 Now glue the two shorter pieces on top of one end of the stack, so that the stack is thicker at one end. This is the neck of the guitar.

8 Mix some glue into the paint. This will thicken it and add strength to the neck when applied to it.

The neck of a guitar has to be strong enough to withstand the stress of the tight strings against it.

9 Apply the glue-and-paint mixture all over your guitar's neck, and leave it for half an hour or so to dry and set.

To make it look like a real guitar, we've decorated ours with painted frets. Frets are metal strips on the neck that help the player find the notes.

This is to ensure that the neck will line up with the side of the tub.

10 Apply a few pieces of duct tape around both ends of the guitar's neck, to strengthen them further.

11 Stand the tub on one of its shorter sides. Hold the thick end of the neck against the rim of the tub. Make a mark on the tub next to the thinner part of the neck.

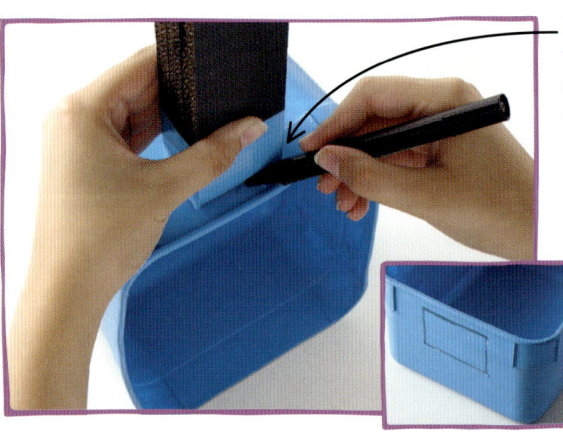

Try to position the neck in the middle of the tub's side.

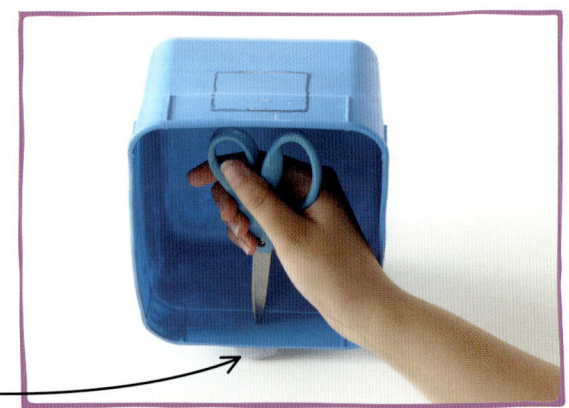

Put adhesive putty under the tub to protect the table.

12 Hold the thin end of the neck against the end of the tub, and line up the top of it with the mark you made. Draw around the neck.

13 Repeat steps 11 and 12 at the other end of the tub. Now carefully use the scissors to make a hole in the middle of each rectangle, from the inside of the box outwards.

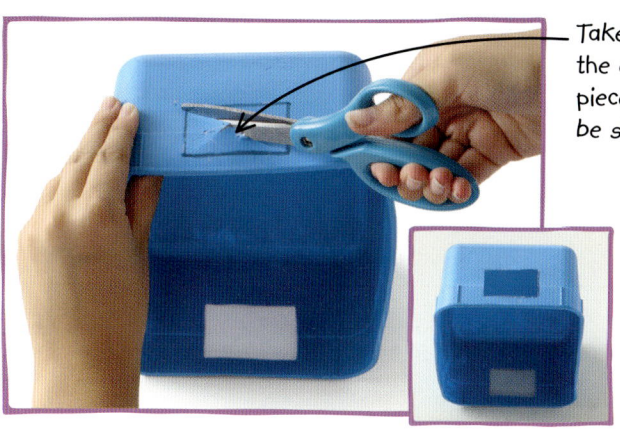

Take care as the edges of the pieces will be sharp.

14 Neatly cut out each rectangle, starting at the hole you made. Cut straight lines out to the corners first, then along the rectangle's sides.

If the neck doesn't fit, you might need to make the hole a bit bigger.

15 Push the thin end of the neck through the two holes, until the thick part of the neck juts up against the side of the tub.

Make two of the holes closer to the end than the other two.

16 Line up a ruler at the thick end of the neck. Mark dots at 1 cm (½ in) intervals, making two of them closer to the end, as shown.

17 Screw the four hooks into each of the holes. These will hold the guitar's strings.

19 Screw a hook into each of the four dots you drew. In each case, make sure the open part of the hook faces away from the body of the guitar.

The box will act as a soundboard, or soundbox, amplifying the sound made by the strings.

18 Draw four more dots on the tape at the thin end of the neck. Make these 1 cm (½ in) apart, too, but all in a line this time.

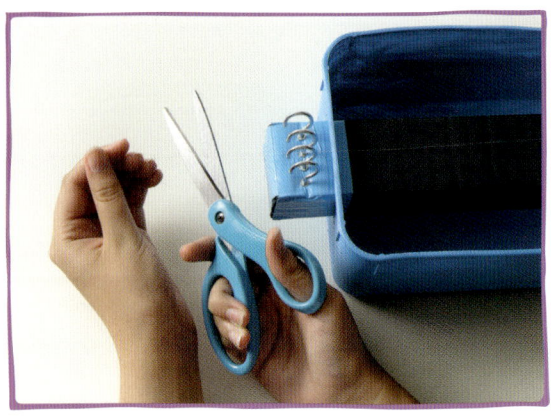

20 Cut four pieces of fishing line, 10 cm (4 in) longer than the distance from one set of hooks to the other set of hooks.

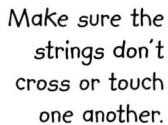

Make sure the strings don't cross or touch one another.

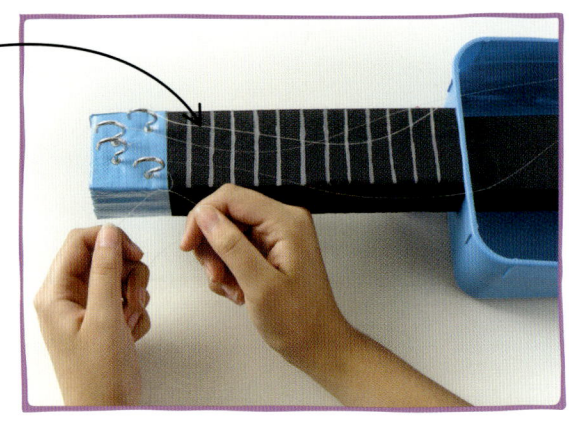

21 At the thick end of the neck, attach each length of fishing line to a hook. Tie a double knot as tightly as you can.

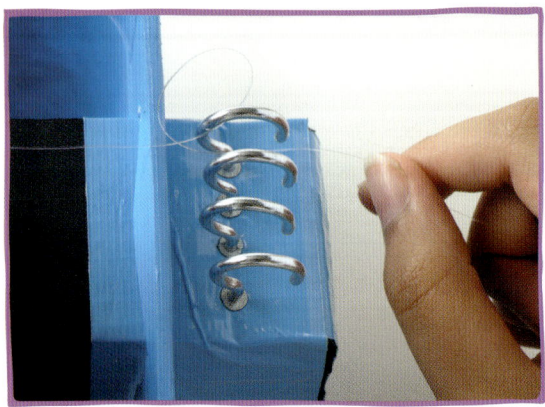

22 Pull the free end of each length taut and wrap it once or twice around a hook at the other end. Do not tie the line at this end.

23 Pull the ends of the four lengths of fishing line together, keeping them taut, and then secure them with the bulldog clip. See what happens when you move your fingers up and down the guitar neck while plucking the strings at the same time.

Press down on one or more strings with the fingers of one hand.

Pluck one or more strings with the thumb or fingers of your other hand.

The body of the guitar is what amplifies the sound.

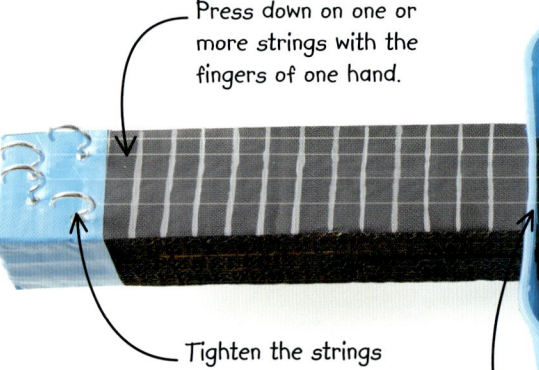

Tighten the strings by turning the hooks at this end until they are taut.

There should be a gap between the strings and the rim of the tub.

You can trim the ends of the strings, but don't make them too short.

HOW IT WORKS

Plucking a string on your guitar causes the string to vibrate many times per second. The more tension there is in the string (the more tightly it's pulled), the more rapidly it vibrates – and the higher the pitch. Pressing a string also raises the pitch. When you press a string, it touches the guitar body in the middle and only its lower half vibrates. This produces a note one octave higher. The strings cause the body of the guitar to vibrate, which disturbs much more air, because it has a larger surface area. This amplifies the vibration of the strings, making it louder.

Plucking the string causes it to vibrate.

Pressing the strings down with your fingers on the cardboard changes the length of the string, which changes the note.

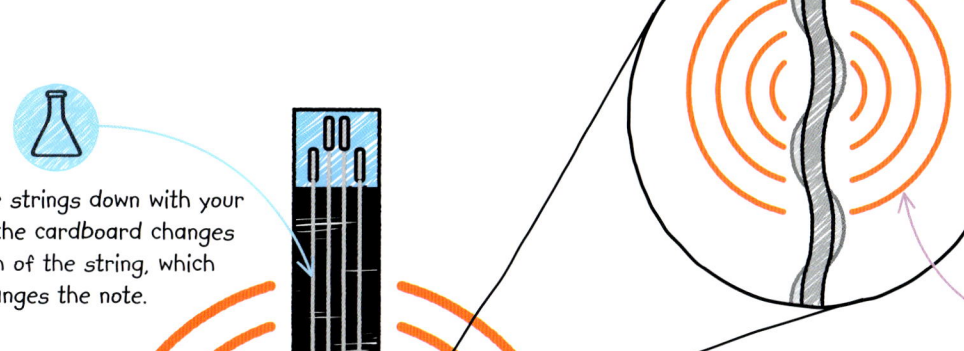

As the body of the guitar vibrates, it amplifies the sound of the strings.

The length, tension, and thickness of the string all affect the note produced.

Without the body of the guitar, the sound waves created by the string's vibrations would barely be audible.

REAL WORLD: TECHNOLOGY
ACOUSTIC GUITAR

Most acoustic guitars have six strings, with each one being a different thickness. The thicker the string, the lower the note it makes. This allows a guitar to produce a great range of notes and sound. Also, unlike your ice-cream-tub guitar, acoustic guitars have a closed front part, with a sound hole, which helps to amplify deeper sounds, as the air inside the guitar is compressed and expands. Finally, the material the guitar is made from greatly affects the sound, as certain materials produce different kinds of sound. Though wood is the most common material, acoustic guitars can also be made from metal or plastic.

GLOSSARY

ACID
A *substance* that has a pH of less than 7. Strong acids (with pH *between* 1 and 3) can burn your *skin*. Weak acids (with pH *between* 4 and 7) are present in vinegar, lemon juice, and cola.

AERODYNAMICS
The study of how air moves around objects, and how the air produces forces such as air resistance and lift.

AIR RESISTANCE
A force that slows down moving objects as they travel through air.

ATMOSPHERIC PRESSURE
The pressure of the air around you, also known as air pressure.

ATOM
A tiny particle of matter. An atom is the smallest part of an element that can exist.

BASE
A *substance* that has a pH of more than 7. A *base* is the chemical opposite of an acid.

BEARING
Part of a machine that reduces friction *between* moving parts. There are bearings in wheels, for example, that allow the wheel to spin freely.

CAM
A machine part that turns rotation into back-and-forth or up-and-down motion.

CARBON DIOXIDE
A chemical compound that is found as a gas in the atmosphere and in fizzy drinks.

CHEMICAL
A substance that is the same all the way through – it is not a mixture. Chemicals can be elements or compounds, and may be liquids, solids, or gases.

CHEMICAL REACTION
A process in which the atoms of two or more chemicals interact to make new chemicals.

COMPOUND
A chemical made of two or more elements. Water is a compound made of the elements hydrogen and oxygen.

COMPRESSION
A squashing force, the opposite of tension.

CONDUCTION
The flow of heat or electricity through a material.

CONDUCTOR
A material through which heat or electricity flows easily. Metals are good conductors.

CRANK
A machine part that can turn rotation into to-and-fro motion or do the opposite.

CYLINDER
A three-dimensional shape that has a circle as its cross-section.

DENSITY
A measure of how much mass is present in a certain volume of a *substance*.

ELECTRON
A negatively charged particle found in atoms. Electricity is a flow of electrons.

ELEMENT
A *substance* made of just one type of atom that cannot be broken down into a simpler substance by chemical reactions.

ELLIPSE
An oval, or flattened circle. The orbits of planets around the Sun are elliptical.

ENERGY
The ability to make things happen. Energy can take various forms, such as electrical energy, kinetic energy (the energy of moving objects), and potential energy (stored energy).

EVAPORATION
The process by which a liquid turns into a gas.

FLUTTER
An energetic vibration created as an object moves through the air (or as air moves past it). The forces exerted by the air cause the object to turn one way, then the other.

FORCE
A push or a pull. Forces change how an *object* moves: by causing it to start or stop moving, speed up or slow down, or change direction. Forces can also change the shape of an object.

FRICTION
A force between surfaces that are in contact. Friction *between* a tyre and the ground pushes a bicycle along as the wheels turn.

GEL
A mixture in which tiny drops of liquid are held in a solid. Jelly is a gel. A gelling agent is a *substance* that is added to water to turn it into a gel.

GENERATOR
A device that produces electricity when it spins around.

GRAVITY
A force that pulls objects together. Earth's gravity pulls things towards the ground.

HYDROGEN ION

A hydrogen atom that has either lost or gained an electron. The more hydrogen ions in a solution, the lower the solution's pH. Acids release lots of hydrogen ions when they dissolve in water, and bases gain them.

INSULATOR

A material through which heat passes slowly. Your clothes insulate you, slowing down the loss of your body heat to the air around you.

ION

An atom that has a negative or positive electric charge.

LEVER

A rigid bar that modifies force or motion when it swings around a fixed point known as a pivot.

LIFT

An upward force on an object moving through the air. It is the result of the air pressure being greater beneath the object than it is above.

MASS

A measure of the amount of matter (stuff) in an object. The force of gravity pulls on everything with mass, so the more mass something has, the more it weighs.

MIXTURE

A substance made of two or more compounds or elements. A mixture can be composed of solids, liquids, and gases. Air is a mixture of gases.

MOLECULE

Two or more atoms joined together.

ORBIT

The path of a planet, comet or asteroid, around the Sun – or the path of a moon or a satellite around a planet. The force of gravity keeps objects in their orbits.

pH

A measure of the concentration of hydrogen ions in a solution. The more hydrogen ions, the lower the pH, and the more acidic the solution.

PIGMENT

A colourful substance. Inks, paints, and flowers all contain pigments.

PRESSURE

A measure of how much a force pushes on a surface.

PYRAMID

A three-dimensional shape with a point at the top and a triangle or a square at the base.

RADIATION

The loss of heat from a hot object (as it gives out infrared radiation). Also short for electromagnetic radiation. Light, infrared, ultraviolet, radio waves, and X-rays are all forms of electromagnetic radiation.

RECYCLING

The process of reusing something that is no longer needed.

SOLUTION

A mixture where one substance is dissolved in a liquid.

SOUND WAVE

An invisible wave that travels through air (or through liquids and solids) as alternating zones of high and low pressure.

SPECTRUM

A spread of colours produced by splitting light into the colours of which it is made, as happens in a rainbow.

TENSION

A pulling force, the opposite of compression.

TURBINE

A device with rotating fan blades that are driven by the pressure of gases, liquids, or steam. Turbines powered by the wind or by moving water are often used to generate electricity.

VIBRATION

A very rapid back-and-forth movement. Guitar strings vibrate when you pluck them, creating sound waves.

VOLUME

The amount of space something takes up, normally measured in millilitres, litres, or cubic metres.

WAVELENGTH

The distance between two peaks of a wave. In a sound wave, the wavelength is the distance between one point of highest air pressure and the next.

WEIGHT

The downward force on an object caused by gravity. The more mass something has, the more it weighs.

INDEX

ACKNOWLEDGMENTS

The publisher would like to thank the following people for their assistance in the preparation of this book:
Sam Atkinson and Pauline Savage for editorial assistance; Smiljka Surla for design assistance; Steve Crozier and Adam Brackenbury for picture retouching; Pankaj Sharma, Ashok Kumar, Nityanand Kumar, and Jagtar Singh for repro work; Sean T. Ross for testing the experiments; Clarisse Hassan for additional illustrations; Helen Peters for indexing; Victoria Pyke for proofreading; Emmie-Mae Avery, Amelia Collins, Lex Hebblethwaite, Mollie Penfold, Melissa Sinclair, Kelly Wray, Abi Wright for hand modelling.

The publisher would like to thank the following for their kind permission to reproduce their photographs:
(Key: a-above, b-below/bottom, c-centre, f-far, l-left, r-right, t-top)
17 Dreamstime.com: Masezdromaderi (tr); Getty Images: Stringer / Bill Pugliano / Getty Images News (cr). 23 Getty Images: Jeff Rotman / The Image Bank (bl). 29 iStockphoto.com: BlackJack3D (crb). 37 Dreamstime.com: Toldiu74 (bl). 41 Alamy Stock Photo: Michele and Tom Grimm (crb). 49 Dreamstime.com: Andrey Shupilo (br). 59 Dreamstime.com: Vladislav Kochelaevskiy (br). 65 iStockphoto.com: oversnap (bl). 69 Depositphotos Inc: alexlmx (br). 77 Dorling Kindersley: Stephen Oliver (crb). 81 Dreamstime.com: Andrey Armyagov (bl). 93 Dreamstime.com: Jarrun Klinmontha (crb). 97 Dorling Kindersley: Natural History Museum, London / Harry

Taylor (cb); Dreamstime.com: Horseman 82 (clb); Science Photo Library: Steve Lowry (crb). 105 123RF.com: Songquan Deng (crb); Dreamstime.com: Ian Klein (clb). 111 Science Photo Library: Laguna Design (bl). 117 Dreamstime.com: Hayati Kayhan (br). 123 The Reinforced Earth Company: (br). 131 Dreamstime.com: STRINGERimages (cra). 137 NASA: ESA / Hubble & NASA (bl). 141 Getty Images: Inti St Clair / Blend Images (bl). 145 Dreamstime.com: Elitsa Lambova (br). 149 Rex by Shutterstock: AP (bc). 157 Dreamstime.com: Mrchan (br).

All other images © Dorling Kindersley Limited.

DK WHAT WILL YOU MAKE NEXT?

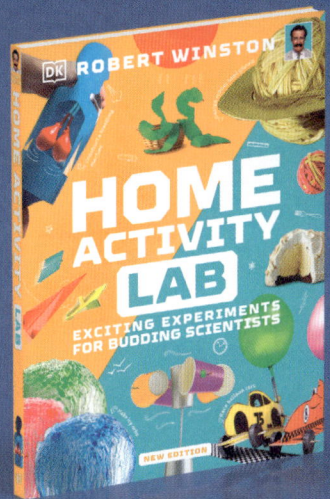

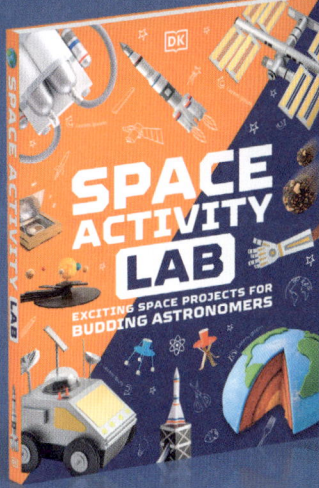

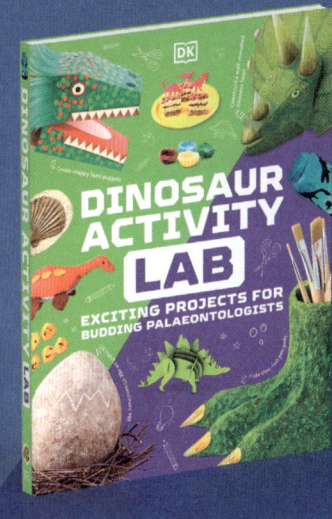

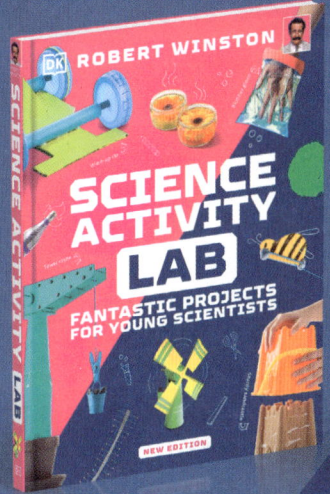

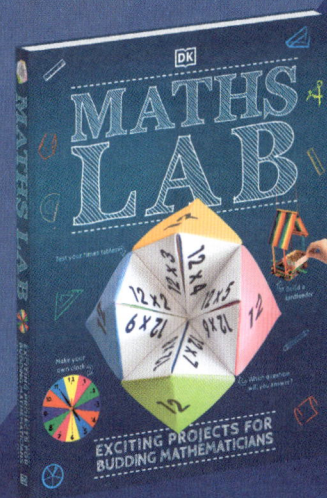

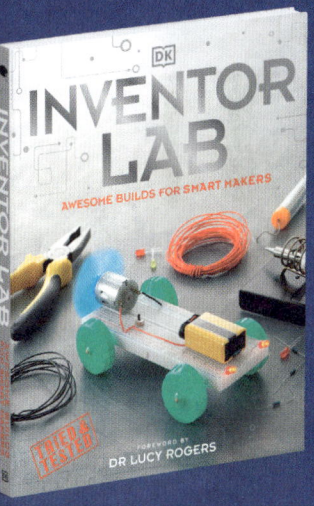

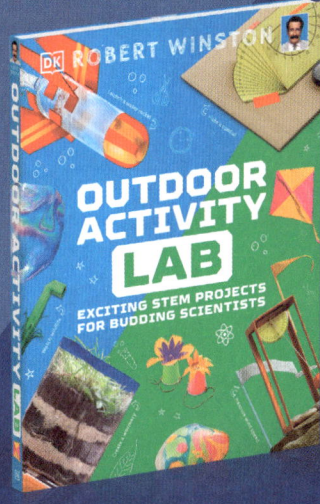

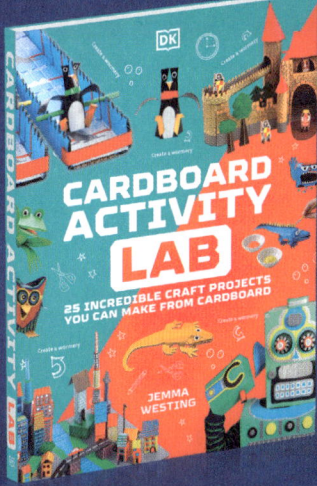

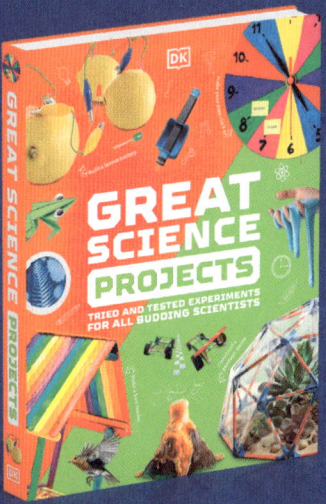